The SMP Concept-based 3D Constitutive Models for Geomaterials

BALKEMA - Proceedings and Monographs
in Engineering, Water, and Earth Sciences

The SMP Concept-based 3D Constitutive Models for Geomaterials

Hajime Matsuoka

Department of Civil Engineering,
Nagoya Institute of Technology, Japan

De'an Sun

Department of Civil Engineering,
Shanghai University, China

CRC Press
Taylor & Francis Group
Boca Raton London New York

CRC Press is an imprint of the
Taylor & Francis Group, an **informa** business
A TAYLOR & FRANCIS BOOK

CRC Press
Taylor & Francis Group
6000 Broken Sound Parkway NW, Suite 300
Boca Raton, FL 33487-2742

First issued in hardback 2019

Typeset in Times New Roman by
Newgen Imaging Systems (P) Ltd, Chennai, India

ISBN-13: 978–0–415–39504–5 (hbk)

British Library Cataloguing in Publication Data
A catalogue record for this book is available from the British Library

Library of Congress Cataloging in Publication Data
Matsuoka, Hajime, 1943–
The SMP concept-based 3D constitutive models
for geomaterials / Hajime Matsuoka, De'an Sun.
p. cm.
Includes bibliographical references and index.
ISBN 0–415–39504–6 (hardcover: alk. paper) 1. Soil mechanics–
Mathematical models. I. Sun, De'an, 1962– II. Title.

TA710.A1.M37 2006
624.1'5136015118–dc22 2005035141

Visit the Taylor & Francis Web site at
http://www.taylorandfrancis.com

and the CRC Press Web site at
http://www.crcpress.com

Contents

Figures

Tables

Preface

This book describes the SMP (Spatially Mobilized Plane) concept, the way to combine the SMP criterion with the well-known Cam-clay model, and the application of the SMP criterion to the elastoplastic constitutive models for geomaterials including clays, sands, and unsaturated soils.

The first author has been studying the mechanics of granular materials such as sands from the particle-level view point from the early 1970s. Especially, he investigated in which part (or zone) of the specimen the soil particles were most mobilized under the two-dimensional and the three-dimensional stress states. This microscopic study provided valuable insight into the shearing mechanism of granular soils in the three-dimensional (3D) state. Based on this study, a three-dimensional failure criterion for soils was proposed by Matsuoka and Nakai (1974), which was named SMP criterion, and it is now recognized worldwide. The SMP criterion is considered to be the three-dimensional equivalent of the two-dimensioned Mohr–Coulomb criterion. Chapter 1 reviews the SMP concept and its application to the prediction of deformation and strength of soils in the three-dimensional stresses.

As well known, the Cam-clay model is an elastoplastic constitutive model for normally consolidated clays, and is now becoming a fundamental constitutive model in the field of geotechnical engineering. Although many books introduce the Cam-clay model, we here try to explain it as easily as possible, in particular from the point of view of how it can predict strains. Chapter 2 simply introduces the original and the modified Cam-clay models.

The Cam-clay model is essentially suitable for clays under the triaxial compression stress state while the SMP criterion is one of the yield and failure criteria for soils in the three-dimensional stresses. Therefore, it is natural and necessary to integrate the Cam-clay model with the SMP criterion. We performed this by introducing a "transformed stress," which is obtained from a transformation of the SMP criterion to a cone in the "transformed" principal stress space. Chapter 3 deals with the method to combine the Cam-clay model with the SMP criterion and shows its performance in the three-dimensional stress state.

The models described in Chapter 3 are essentially applicable to the normally consolidated clays under the general stress state, and they can not accurately predict the stress–strain behavior of other geomaterials such as sands and unsaturated soils. In Chapter 4 we further introduce three elastoplastic constitutive models for sands, anisotropically consolidated soils, and unsaturated soils. All models have been developed in the three-dimensional stresses by using the SMP criterion.

The elastoplastic constitutive tensors of all these introduced models are given for readers to apply the models to engineering practice by the finite element method.

Hajime Matsuoka is grateful to the late Prof. S. Murayama (former Professor Emeritus of Kyoto University) for teaching him the importance of studying the origin of phenomena. A grateful acknowledgment is made to the late Prof. T. Mogami (former Professor Emeritus of the University of Tokyo), Prof. M. Satake (Professor Emeritus of Tohoku University), and Prof. D. Karube (Professor Emeritus of Kobe University) for their helpful advice and encouragement given throughout this study.

The authors are most appreciative of the contributions made by Prof. Y. P. Yao at Beijing University of Aeronautics and Astronautics, China, who worked out the transformed stress and the hardening parameter with the authors during his stay at the Nagoya Institute of Technology (NIT), Japan. The authors also sincerely thank Prof. T. Nakai at Nagoya Institute of Technology for making the study of the SMP criterion together with the authors and using his experimental data in this book.

Many students have been particularly helpful in conducting the laboratory experiments and modeling soil behavior. They are M. Ando, A. Kogane, N. Fukuzawa, H. Ishii, W. Ichihara, T. Kinoshita, K. Tani, and H. Honda.

Very special thanks go to Dr D. C. Sheng at the University of Newcastle, Australia, for the thorough reviews of the English language and text of the book.

Chapter 1

Spatially mobilized plane (SMP) and SMP criterion

1.1 Origin of SMP

In order to understand the shear mechanism of granular materials such as sand and gravel from microscopic point of view, direct shear tests (Figs 1.1 and 1.2) and biaxial compression tests (Figs 1.3 and 1.4) have been performed using assemblies of aluminum rods and photoelastic rods. Here, the aluminum rods and photoelastic rods are used as two-dimensional models of granular materials. It can be seen from the particle movement in the figures that there is a special plane or a shear band which plays a dominant role in controlling the deformation and failure of the granular material during shearing. The potential slip plane is a plane between the upper and lower boxes in direct shear tests (Figs 1.1 and 1.2) and a plane inclined at $45^\circ + \phi_{mo}/2$ (ϕ_{mo} = mobilized internal friction angle) to the major principal stress plane (usually horizontal plane). This plane is called the mobilized plane by Matsuoka (1974b). This mobilized plane becomes noticeable at failure (from peak to residual), and is usually called the slip plane. If the plane plays a dominant role in controlling the failure behavior of granular materials, it is natural to assume that the plane also plays a dominant role in controlling the deformation behavior up to failure. For example, as shown in Figure 1.2, the relative movement of the upper and lower particles along the central horizontal plane is largest within the specimen, and the plane can be considered to control the deformation and strength of the specimen. The shear resistance is the summation of the horizontal components of the contact forces between particles across the mobilized plane. Therefore, the shear resistance depends on the contact forces, the contact angles, and the mobilized friction angles between particles across the mobilized plane (Matsuoka 1974a).

The assumption that there exists a mobilized plane controlling deformation and failure of granular materials has the following merit. The results of apparently different shear tests, such as direct shear test, simple shear test, triaxial compression test, triaxial extension test, plane strain test, and true triaxial test, can be arranged according to the same stress–strain relationship, which will be explained later. That is to say, the shear behavior of granular materials in different shear tests are essentially the same if the concept of the mobilized plane is used.

1.2 Stress–strain relation based on compositely mobilized plane (CMP)

The discussions in the previous section are based on the two-dimensional behavior of soil particles. However, real soil particles exhibit three-dimensional (3D) behavior. Let us discuss how to determine the mobilized plane in this case.

Figure 1.1 Direct shear test on an assembly of aluminum rods.

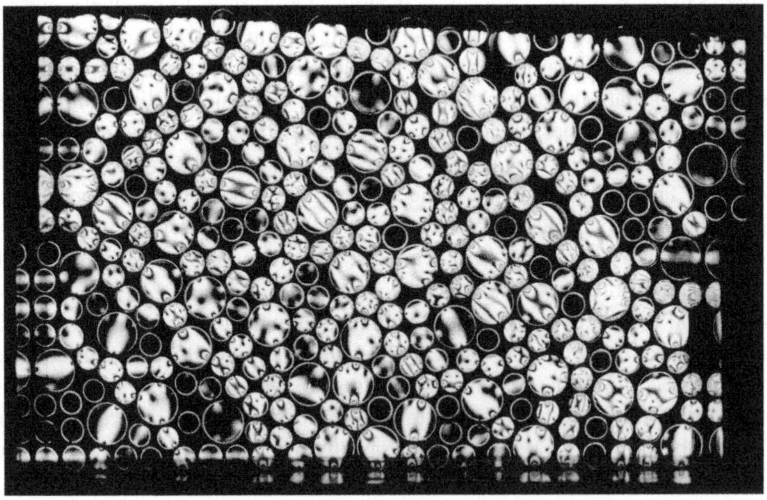

Figure 1.2 Direct shear test on an assembly of photoelastic rods.

Figure 1.5 shows three Mohr's stress circles under respective two principal stresses. The three points (P_1, P_2, and P_3) at which three straight lines from the origin are tangent to the three Mohr's circles correspond to the stress states of the three two-dimensional mobilized planes in which the shear-normal stress ratio is maximum. If the mechanical behavior

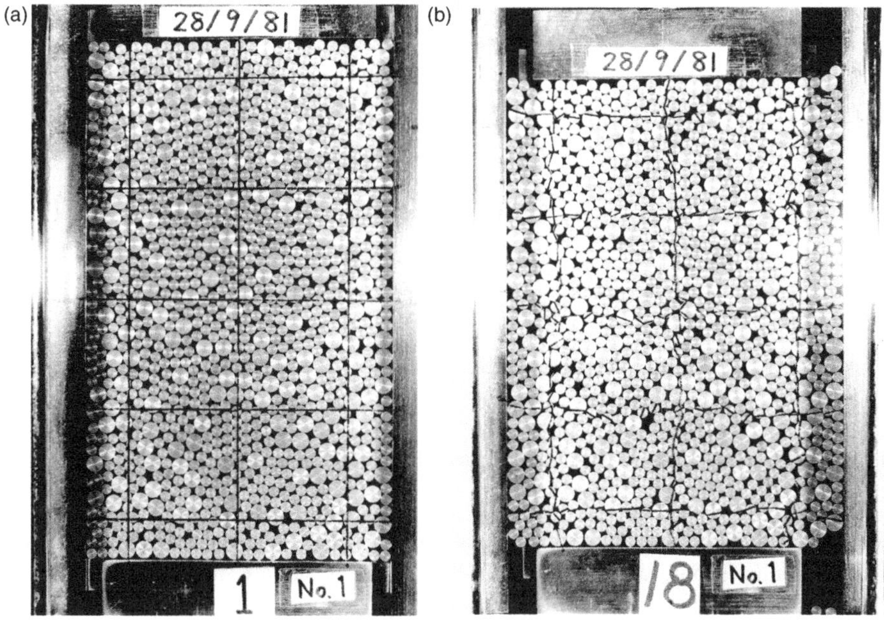

Figure 1.3 Biaxial compression test on an assembly of aluminum rods: (a) initial state and (b) at failure.

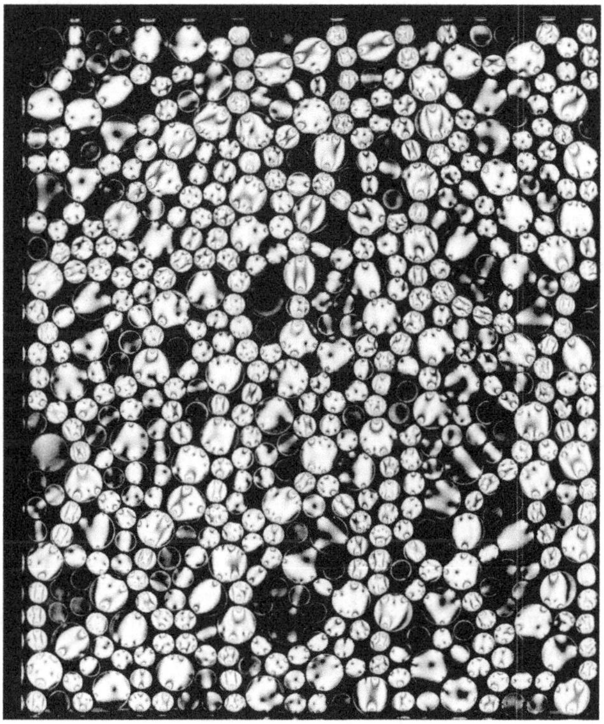

Figure 1.4 Biaxial compression test on an assembly of photoelastic rods.

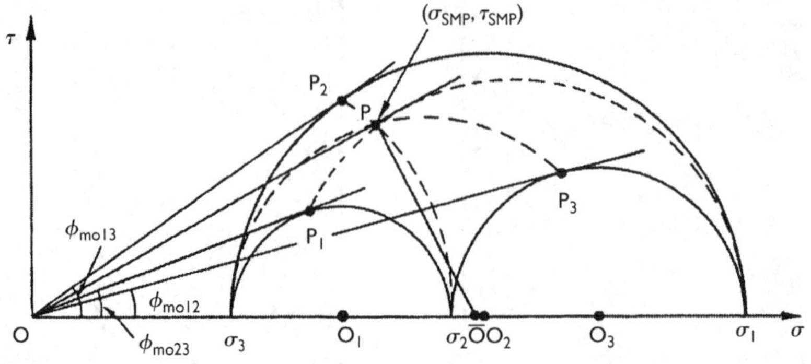

Figure 1.5 Normal and shear stresses on mobilized planes and SMP (Satake 1978).

of soil particles is governed by the frictional law, the plane where the shear-normal stress ratio is maximum should correspond to the soil particles that are most mobilized on average (Murayama 1966). When three different principal stresses σ_1, σ_2, and σ_3 are applied to three orthogonal directions, three two-dimensional mobilized planes between the respective two principal stress directions results shown in Figure 1.6(a). These three two-dimensional mobilized planes AB, BC, and AC are called the Compositely Mobilized Planes (CMP) by Matsuoka (1974b).

Figure 1.7 shows "two" two-dimensional mobilized planes in triaxial compression stress state ($\sigma_1 > \sigma_2 = \sigma_3$) and triaxial extension stress state ($\sigma_1 = \sigma_2 > \sigma_3$), respectively. In these cases, there are only "two" mobilized planes because the mobilized plane does not occur when two principal stresses are identical.

Figure 1.8 shows the shapes of sand specimens after triaxial compression and triaxial extension tests. The marked lines were horizontal or vertical before tests. It can be seen from Figure 1.8 that mobilized planes similar to those shown in Figure 1.7 exist in the specimens.

Figure 1.9 shows "two-dimensional" principal strain increments under respective pairs of principal stresses. In the figure, $d\varepsilon_{1(12)}$ and $d\varepsilon_{2(12)}$ denote two-dimensional principal strain increments under σ_1 and σ_2. Similarly, $d\varepsilon_{2(23)}$ and $d\varepsilon_{3(23)}$ denote two-dimensional principal strain increments under σ_2 and σ_3, and $d\varepsilon_{3(13)}$ and $d\varepsilon_{1(13)}$ denote two-dimensional principal strain increments under σ_1 and σ_3. Under three principal stresses σ_1, σ_2, and σ_3, the principal strain increment $d\varepsilon_1$, $d\varepsilon_2$, and $d\varepsilon_3$ may be expressed by assuming the superposition of these "two-dimensional" principal strain increments as follows (Matsuoka 1974b):

$$\left.\begin{aligned} d\varepsilon_1 &= d\varepsilon_{1(12)} + d\varepsilon_{1(13)} \\ d\varepsilon_2 &= d\varepsilon_{2(12)} + d\varepsilon_{2(23)} \\ d\varepsilon_3 &= d\varepsilon_{3(23)} + d\varepsilon_{3(13)} \end{aligned}\right\} \tag{1.1}$$

In the case of triaxial compression condition ($\sigma_1 > \sigma_2 = \sigma_3$), the principal strain increments are expressed as follows:

$$\left.\begin{aligned} d\varepsilon_1 &= 2 \cdot d\varepsilon_{1(13)} \\ d\varepsilon_2 &= d\varepsilon_3 = d\varepsilon_{3(13)} + d\varepsilon_{3(33)} \end{aligned}\right\} \tag{1.2}$$

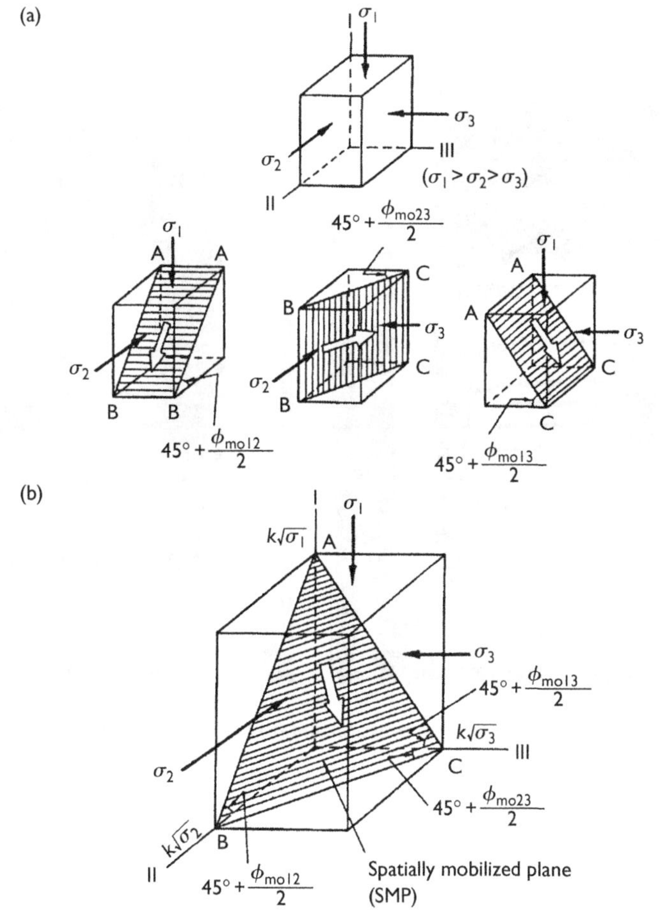

Figure 1.6 Geometrical representations of mobilized planes and SMP in three-dimensional space.

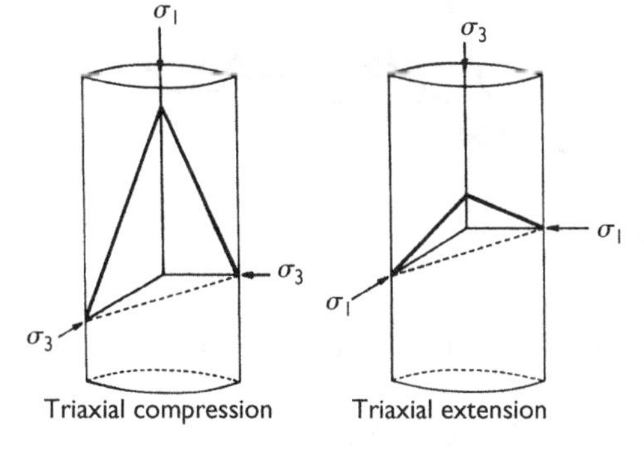

Figure 1.7 "Two" two-dimensional mobilized planes under triaxial compression and triaxial extension conditions.

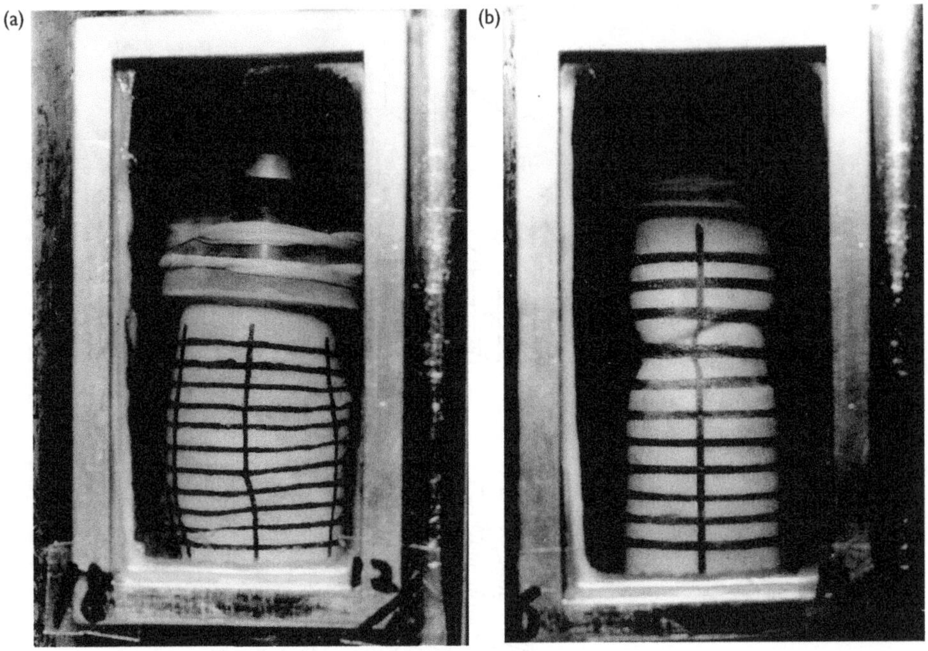

Figure 1.8 Deformation of (a) triaxial compression and (b) triaxial extension specimens after tests.

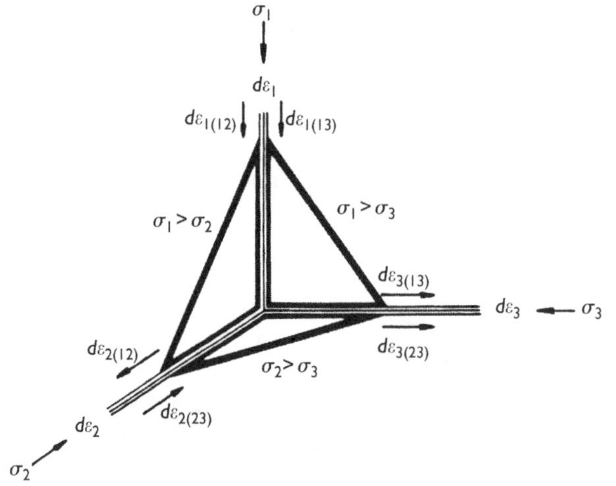

Figure 1.9 "Two-dimensional" principal strain increments under respective pairs of principal stresses.

Similarly, in the case of triaxial extension condition ($\sigma_1 = \sigma_2 > \sigma_3$), the principal strain increments become

$$\left.\begin{aligned} d\varepsilon_1 &= d\varepsilon_2 = d\varepsilon_{1(13)} + d\varepsilon_{1(11)} \\ d\varepsilon_3 &= 2 \cdot d\varepsilon_{3(13)} \end{aligned}\right\} \tag{1.3}$$

where $d\varepsilon_{3(33)}$ in Eq. (1.2) and $d\varepsilon_{1(11)}$ in Eq. (1.3) correspond to the "two-dimensional" strain increments under the two conjugate principal stresses. Therefore, the following equation for isotropic compression can be obtained by assuming a linear *e*-log *p* relation.

$$d\varepsilon_{3(33)} = d\varepsilon_{1(11)} = \frac{1}{6}\frac{0.434C_c}{1+e_0}\frac{d\sigma_m}{\sigma_m} \tag{1.4}$$

where σ_m is the effective mean principal stress, C_c is the compression index, and e_0 is the initial void ratio. The coefficient of $\frac{1}{6}$ in Eq. (1.4) is due to the fact that the volumetric strain increment is a summation of the six terms on the right-hand side of Eq. (1.1) (see Fig. 1.9). In the case of isotropic compression, these terms are identical.

To obtain the "two-dimensional" principal strain from the results of triaxial compression tests, it is understood by Eqs (1.2) and (1.4) that the measured ε_1 should be divided by 2 and the principal strain due to isotropic compression (Eq. 1.4) should be subtracted from the measured ε_3. In the case of triaxial extension tests, the measured ε_3 should be divided by 2 and the principal strain due to isotropic compression should be subtracted from the measured ε_1. Under a constant effective mean principal stress σ_m, $d\varepsilon_{3(33)}$ in Eq. (1.2) and $d\varepsilon_{1(11)}$ in Eq. (1.3) become zero as seen from Eq. (1.4). Therefore, in the case of σ_m-constant tests, the "two-dimensional" principal strain increments $d\varepsilon_{1(13)} = d\varepsilon_{1(\text{measured})}/2$ and $d\varepsilon_{3(13)} = d\varepsilon_{3(\text{measured})}$ for triaxial compression, and $d\varepsilon_{1(13)} = d\varepsilon_{1(\text{measured})}$ and $d\varepsilon_{3(13)} = d\varepsilon_{3(\text{measured})}/2$ for triaxial extension (see Fig. 1.7).

If we obtain the "two-dimensional" principal strain increments $d\varepsilon_{1(13)}$ and $d\varepsilon_{3(13)}$ from triaxial compression tests or triaxial extension tests, the shear strain increment $d\gamma$ and the normal strain increment $d\varepsilon_N$ on "one" two-dimensional mobilized plane can be expressed as follows:

$$\begin{aligned}\frac{d\gamma}{2} &= \frac{d\varepsilon_{1(13)} - d\varepsilon_{3(13)}}{2}\sin\{2(45^\circ + \phi_{mo}/2)\} \\ &= \frac{d\varepsilon_{1(13)} - d\varepsilon_{3(13)}}{2}\cos\phi_{mo}\end{aligned} \tag{1.5}$$

$$\begin{aligned}d\varepsilon_N &= \frac{d\varepsilon_{1(13)} + d\varepsilon_{3(13)}}{2} + \frac{d\varepsilon_{1(13)} - d\varepsilon_{3(13)}}{2}\cos t\{2(45^\circ + \phi_{mo}/2)\} \\ &= \frac{d\varepsilon_{1(13)} + d\varepsilon_{3(13)}}{2} - \frac{d\varepsilon_{1(13)} - d\varepsilon_{3(13)}}{2}\sin\phi_{mo}\end{aligned} \tag{1.6}$$

The shear-normal stress ratio τ/σ_N on the mobilized plane is expressed by the following equation (see Fig. 1.10).

$$\frac{\tau}{\sigma_N} = \tan\phi_{mo} = \frac{\sigma_1 - \sigma_3}{2\sqrt{\sigma_1\sigma_3}} = \frac{1}{2}\left(\sqrt{\frac{\sigma_1}{\sigma_3}} - \sqrt{\frac{\sigma_3}{\sigma_1}}\right) \tag{1.7}$$

The merit of transferring the measured stresses and strains to those on the mobilized plane is that results from different shear tests become comparable and can then be described using the same stress–strain relation. This is because the shear mechanisms of granular materials are the same even in different shear tests. In the following, let us validate the assumptions using the results of triaxial compression tests and triaxial extension tests on sand.

Figure 1.11 shows the relation between the shear-normal stress ratio τ/σ_N and the normal–shear strain increment ratio $(-d\varepsilon_N/d\gamma)$ on "one" two-dimensional mobilized plane.

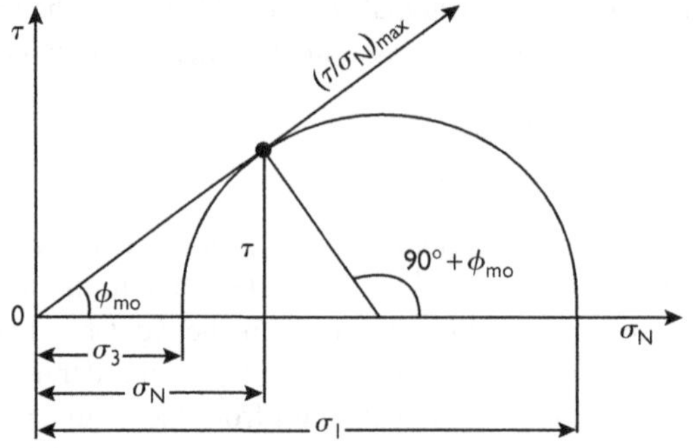

Figure 1.10 A plane where shear–normal stress ratio is maximum.

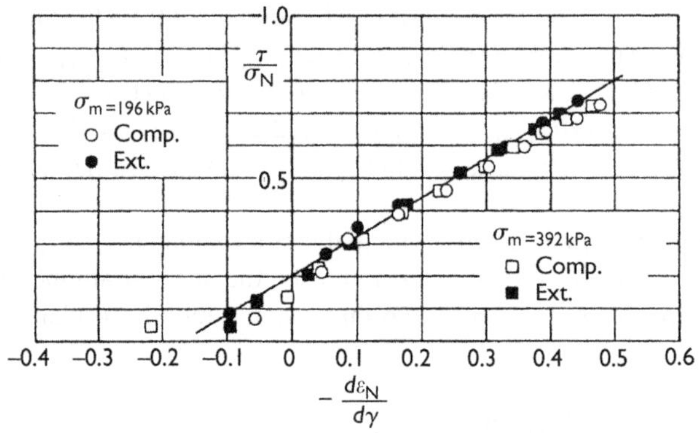

Figure 1.11 Relation between τ/σ_N and $-d\varepsilon_N/d\gamma$ on one mobilized plane obtained by triaxial compression and extension tests on Toyoura sand.

This relation is the so-called stress–dilatancy relation (Rowe 1962). It is very interesting to note that a unique relation between the stress ratio and the strain increment ratio exists for Toyoura sand tested under triaxial compression and triaxial extension conditions with two sets of effective mean stresses (Matsuoka 1983). The unique relation between the stress ratio and the strain increment ratio is expressed as follows (Matsuoka 1974b, 1983):

$$\frac{\tau}{\sigma_N} = \lambda\left(-\frac{d\varepsilon_N}{d\gamma}\right) + \mu \tag{1.8}$$

where λ and μ are soil parameters. This relation corresponds to the flow rule in the theory of plasticity, i.e. the plastic potential can be derived from this unique relation.

Figure 1.12 shows the relations between the shear-normal stress ratio τ/σ_N and the shear strain γ and between the normal strain ε_N and the shear strain γ on "one" two-dimensional mobilized plane (Matsuoka 1983). It is also interesting to note that a unique stress ratio-strain relation exists for triaxial compression and triaxial extension stress states. The hardening function under a constant mean principal stress can be derived from this unique stress–strain relation.

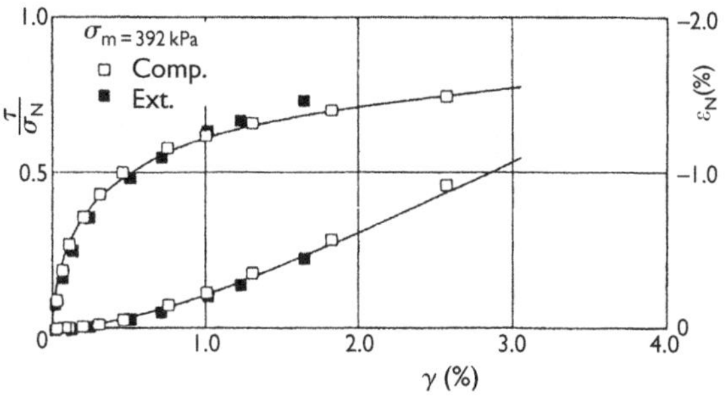

Figure 1.12 Relation between τ/σ_N, γ and ε_N on one mobilized plane obtained by triaxial compression and extension tests on Toyoura sand.

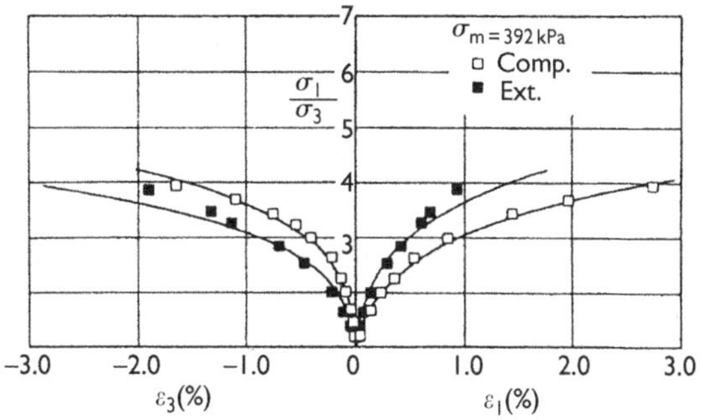

Figure 1.13 Predicted and measured relations between σ_1/σ_3, ε_1 and ε_3 obtained by triaxial compression and extension tests on Toyoura sand.

The two kinds of unique stress–strain relationships shown in Figures 1.11 and 1.12 support the assumption that the stress–strain behavior on "one" two-dimensional mobilized plane is the same if the sample is initially isotropic. Based on such unique stress–strain relations on "one" two-dimensional mobilized plane and the assumed superposition rule according to Eq. (1.1), we can easily calculate the principal strain under three different principal stresses. Figure 1.13 shows the comparison between the measured and calculated principal strains in a triaxial compression test and a triaxial extension test, and the calculated principal strain by the method mentioned here. The marks (□ and ■) are the experimental data and the solid lines are the calculated results.

The uniqueness in stress–strain relations in Figures 1.11 and 1.12 mentioned in the previous paragraph is very important for constructing the constitutive relations for soils. If there is no such uniqueness, we would need to conduct soil tests along all the possible stress paths that will be encountered in practical engineering.

1.3 Stress–strain relation based on SMP

If the three "two-dimensional" mobilized planes AB, BC, and AC in Figure 1.6(a) indeed have the physical meaning as discussed in the previous section, it is natural to think of a plane ABC with the three sides AB, BC, and AC, as shown in Figure 1.6(b). The plane ABC was called Spatially Mobilized Plane (SMP) by Matsuoka and Nakai (1974). The stress state on the SMP can be represented as a point P in Mohr's plane, as shown in Figure 1.5 (Satake 1978). The normal stress σ_{SMP}, the shear stress τ_{SMP}, and the shear-normal stress ratio τ_{SMP}/σ_{SMP} on the SMP are expressed as follows (Matsuoka and Nakai 1974, 1985):

$$\sigma_{SMP} = \sigma_1 a_1^2 + \sigma_2 a_2^2 + \sigma_3 a_3^2 = 3I_3/I_2 \tag{1.9}$$

$$\begin{aligned}\tau_{SMP} &= \sqrt{(\sigma_1 - \sigma_2)^2 a_1^2 a_2^2 + (\sigma_2 - \sigma_3)^2 a_2^2 a_3^2 + (\sigma_3 - \sigma_1)^2 a_3^2 a_1^2} \\ &= \sqrt{I_1 I_2 I_3 - 9I_3^2}/I_2\end{aligned} \tag{1.10}$$

$$\frac{\tau_{SMP}}{\sigma_{SMP}} = \sqrt{\frac{I_1 I_2 - 9I_3}{9I_3}} \tag{1.11}$$

where (a_1, a_2, a_3) represent the direction cosines of the normal to the SMP, and are given by the following equation,

$$a_i = \sqrt{I_3/(\sigma_i I_2)} \quad (i = 1, 2, 3) \tag{1.12}$$

And I_1, I_2, and I_3 are the first, second, and third stress invariants, respectively:

$$\left.\begin{aligned}I_1 &= \sigma_1 + \sigma_2 + \sigma_3 \\ I_2 &= \sigma_1\sigma_2 + \sigma_2\sigma_3 + \sigma_3\sigma_1 \\ I_3 &= \sigma_1\sigma_2\sigma_3\end{aligned}\right\} \tag{1.13}$$

Figure 1.14 shows a vector $\overrightarrow{OP}$ for the principal strain increment $d\vec{\varepsilon}_i = (d\varepsilon_1, d\varepsilon_2$, and $d\varepsilon_3)$ in the principal strain increment space. It is reasonable to consider that the average sliding direction of soil particles coincides with that of the vector of the principal strain increment in the case when the inter-particle contacts are randomly distributed in the soil element. Following this reasoning, we introduce new strain increments based on the SMP, i.e., $d\varepsilon^*_{SMP}$ and $d\gamma^*_{SMP}$, being the normal and tangential components of the principal strain increment vector to the SMP, respectively. Based on the assumption that the principal axes of stress and strain increment tensors are identical, the new strain increments ($d\varepsilon^*_{SMP}$ and $d\gamma^*_{SMP}$), which are represented by the normal component $\overrightarrow{ON}$ and the tangential component $\overrightarrow{NP}$ of the principal strain increment vector $\overrightarrow{OP}$ to the SMP, can be expressed by the following equations (Nakai and Matsuoka 1982, 1983).

$$d\varepsilon^*_{SMP} = d\varepsilon_1 a_1 + d\varepsilon_2 a_2 + d\varepsilon_3 a_3 \tag{1.14}$$

$$d\gamma^*_{SMP} = \sqrt{(d\varepsilon_1 a_2 - d\varepsilon_2 a_1)^2 + (d\varepsilon_2 a_3 - d\varepsilon_3 a_2)^2 + (d\varepsilon_3 a_1 - d\varepsilon_1 a_3)^2} \tag{1.15}$$

where $d\varepsilon^*_{SMP}/d\gamma^*_{SMP}$ is considered to indicate the average sliding direction of soil particles when they are viewed from the SMP, and is expected to have a unique relation with the shear-normal stress ratio τ_{SMP}/σ_{SMP} on the SMP.

Figure 1.15 shows the shear-normal stress ratio τ_{SMP}/σ_{SMP} plotted against the normal-shear strain increment ratio $(-d\varepsilon^*_{SMP}/d\gamma^*_{SMP})$. These results are obtained from triaxial

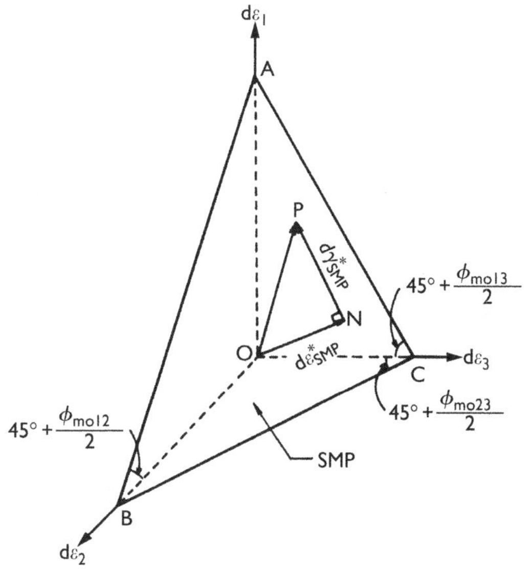

Figure 1.14 Amounts of strain increments $d\varepsilon^*_{SMP}$ and $d\gamma^*_{SMP}$ in principal strain increment space.

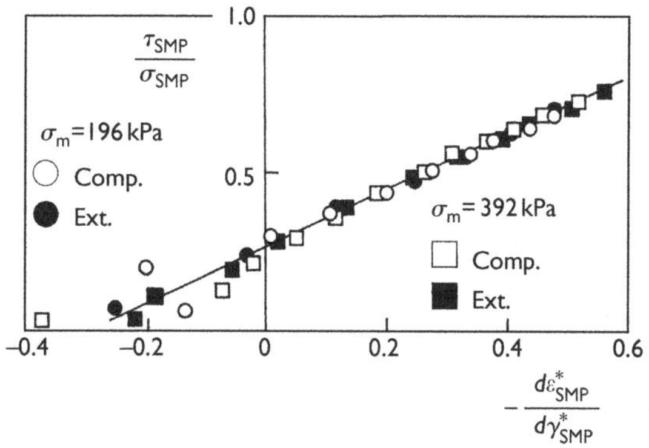

Figure 1.15 Relation between τ_{SMP}/σ_{SMP} and $-d\varepsilon^*_{SMP}/d\gamma^*_{SMP}$ obtained by triaxial compression and extension tests on Toyoura sand.

compression tests and triaxial extension tests on Toyoura sand under two effective mean stresses (Nakai and Matsuoka 1982, 1983). Figure 1.16 presents the same stress ratio versus the strain increment ratio on the SMP, obtained from true triaxial tests on Toyoura sand under a constant effective mean principal stress (Nakai and Matsuoka 1982, 1983). In Figure 1.16, θ denotes the angle from the direction of the major principal stress σ_1 in the π-plane (see Fig. 1.28). It can be seen from Figures 1.15 and 1.16 that a unique relation between the stress ratio τ_{SMP}/σ_{SMP} and the strain increment ratio $(-d\varepsilon^*_{SMP}/d\gamma^*_{SMP})$ is obtained under different principal stresses (σ_m = 196 and 392 kPa). This unique stress ratio versus strain increment

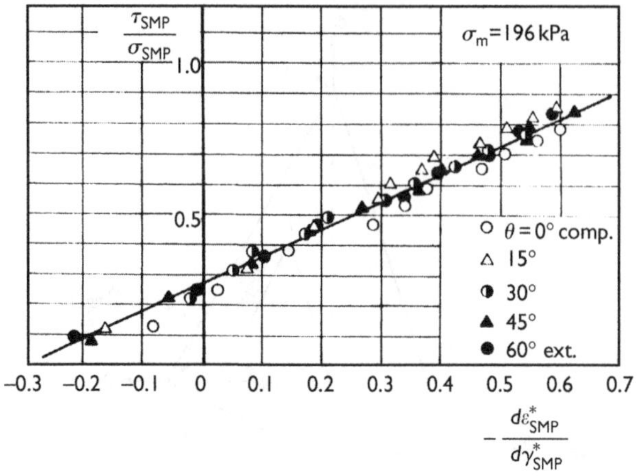

Figure 1.16 Relation between τ_{SMP}/σ_{SMP} and $-d\varepsilon^*_{SMP}/d\gamma^*_{SMP}$ obtained by true triaxial tests on Toyoura sand.

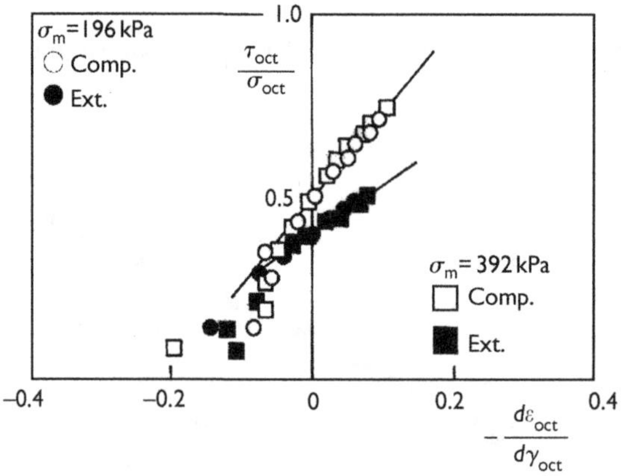

Figure 1.17 Relation between τ_{oct}/σ_{oct} and $-d\varepsilon_{oct}/d\gamma_{oct}$ obtained by the same tests as Figure 1.15.

ratio relation is expressed as follows (Nakai and Matsuoka 1982, 1983):

$$\frac{\tau_{SMP}}{\sigma_{SMP}} = \lambda^* \left(-\frac{d\varepsilon^*_{SMP}}{d\gamma^*_{SMP}} \right) + \mu^* \tag{1.16}$$

where λ^* and μ^* are soil parameters. This relation corresponds to the flow rule in the theory of plasticity, and can be regarded as a stress–dilatancy relation under different principal stresses. In Figure 1.17, the same test results as in Figure 1.15 are rearranged by plotting the shear-normal stress ratio τ_{oct}/σ_{oct} versus the strain increment ratio $(-d\varepsilon_{oct}/d\gamma_{oct})$ on the octahedral plane. It is seen from Figure 1.17 that the points of triaxial compression show a different tendency from the points of triaxial extension.

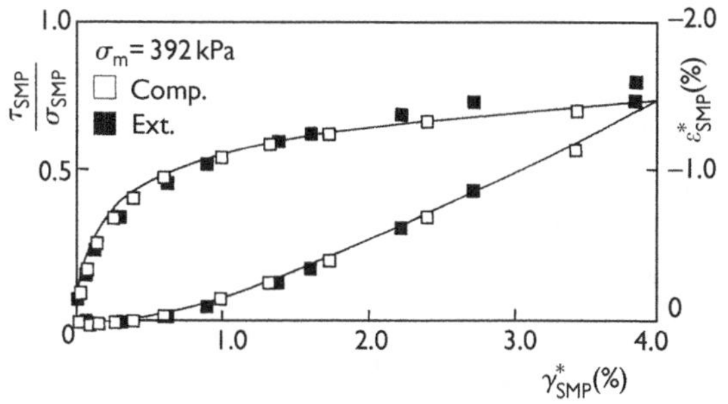

Figure 1.18 Relation between τ_{SMP}/σ_{SMP}, γ^*_{SMP}, and ε^*_{SMP} obtained by triaxial compression and extension tests on Toyoura sand.

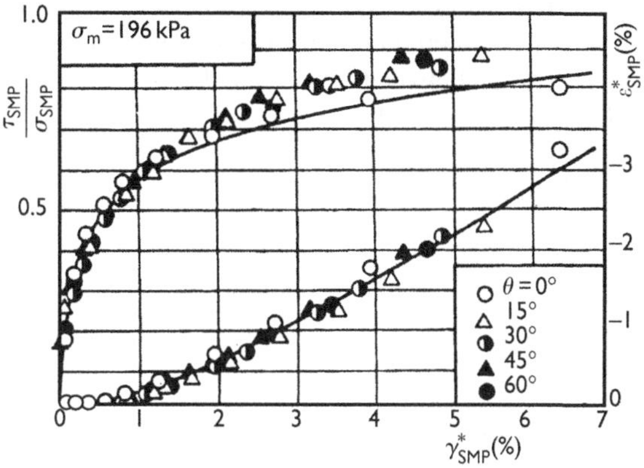

Figure 1.19 Relation between τ_{SMP}/σ_{SMP}, γ^*_{SMP}, and ε^*_{SMP} obtained by true triaxial tests on Toyoura sand.

Figure 1.18 shows the relations between the shear-normal stress ratio τ_{SMP}/σ_{SMP} and the shear strain γ^*_{SMP} and between the normal strain ε^*_{SMP} and the shear strain γ^*_{SMP} on the SMP, obtained from a triaxial compression test and a triaxial extension test on Toyoura sand (Nakai and Matsuoka 1982, 1983). Figure 1.19 presents the same stress ratio-strain relation, but obtained from true triaxial tests on Toyoura sand (Nakai and Matsuoka 1982, 1983). It can be seen from Figures 1.18 and 1.19 that these stress–strain relations are unique under three different principal stresses (σ_m = 196 and 392 kPa). The hardening function under constant effective mean stresses can be derived from the unique stress ratio-strain relation. Such unique stress–strain relations shown in Figures 1.15, 1.16, 1.18, and 1.19 confirm the significance of the SMP: the shear behavior of frictional materials is governed on average by the stress-strain relations on the SMP. In Figure 1.20, the same test results as shown in Figure 1.18 are rearranged by plotting the stress ratio τ_{oct}/σ_{oct} versus strains (γ_{oct} and ε_{oct}) on the octahedral plane. We can see from this figure that the points of triaxial compression show a different tendency from the points of triaxial extension. In many constitutive models for soils, the stress–strain relations in the octahedral plane are assumed to be the same for different stress

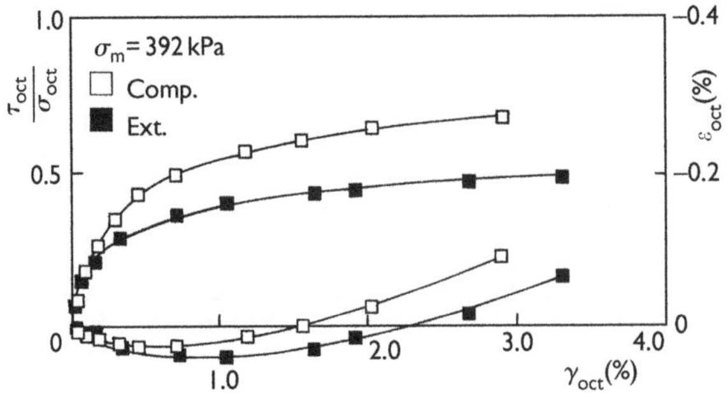

Figure 1.20 Relation between τ_{oct}/σ_{oct}, γ_{oct}, and ε_{oct} obtained by the same tests as Figure 1.18.

states, which clearly conflicts with the experimental data shown in Figures 1.17 and 1.20. In Chapter 3, we will introduce a method to improve this shortcoming.

We have the following relation between the principal strain increments $d\varepsilon_i$ and the strain increments $d\gamma^*_{SMP}$ and $d\varepsilon^*_{SMP}$ on the SMP.

$$d\varepsilon_i = a_i d\varepsilon^*_{SMP} + b_i d\gamma^*_{SMP} \quad (i = 1, 2, \text{ and } 3) \tag{1.17}$$

where

$$b_i = \frac{\sigma_i - \sigma_{SMP}}{\tau_{SMP}} a_i \tag{1.18}$$

Based on the unique stress–strain relations on the SMP, as shown in Figures 1.16 and 1.19, we can easily calculate the principal strains (ε_1, ε_2, and ε_3) under three different principal stresses using Eq. (1.17) together with Eq. (1.18). The soil parameters for the calculation can be determined from the simplest test, that is, triaxial compression test, because of the unique stress–strain relations shown in Figures 1.15, 1.16, 1.18, and 1.19. In Figure 1.21, the measured principal strains from the true triaxial tests ($\theta = 0°$, $15°$, $30°$, $45°$, and $60°$) are compared with the calculated principal strains using the unique stress–strain relations in Figures 1.16 and 1.19 and Eq. (1.17) (Nakai and Matsuoka 1982, 1983).

As mentioned in Section 1.2 and this section, the two-dimensional "mobilized planes" and the SMP play an important role in obtaining the unique stress–strain relations of soils during shearing. The reason is that the deformation and failure behavior of granular materials such as soils are essentially the same when described on the two-dimensional mobilized plane or on the spatially mobilized plane (SMP).

1.4 Failure criteria for metals and granular materials

Well-known failure criteria include the Tresca, the Mises, and the Mohr-Coulomb. The Tresca and Mises criteria are usually adopted for metals, whereas the Mohr-Coulomb criterion is adopted more often for granular materials such as soils. Here, the Matsuoka-Nakai criterion based on the SMP is introduced, which is an extension of the Mohr-Coulomb criterion to three dimensions (Matsuoka and Nakai 1974, 1985; Matsuoka 1976). The relationships

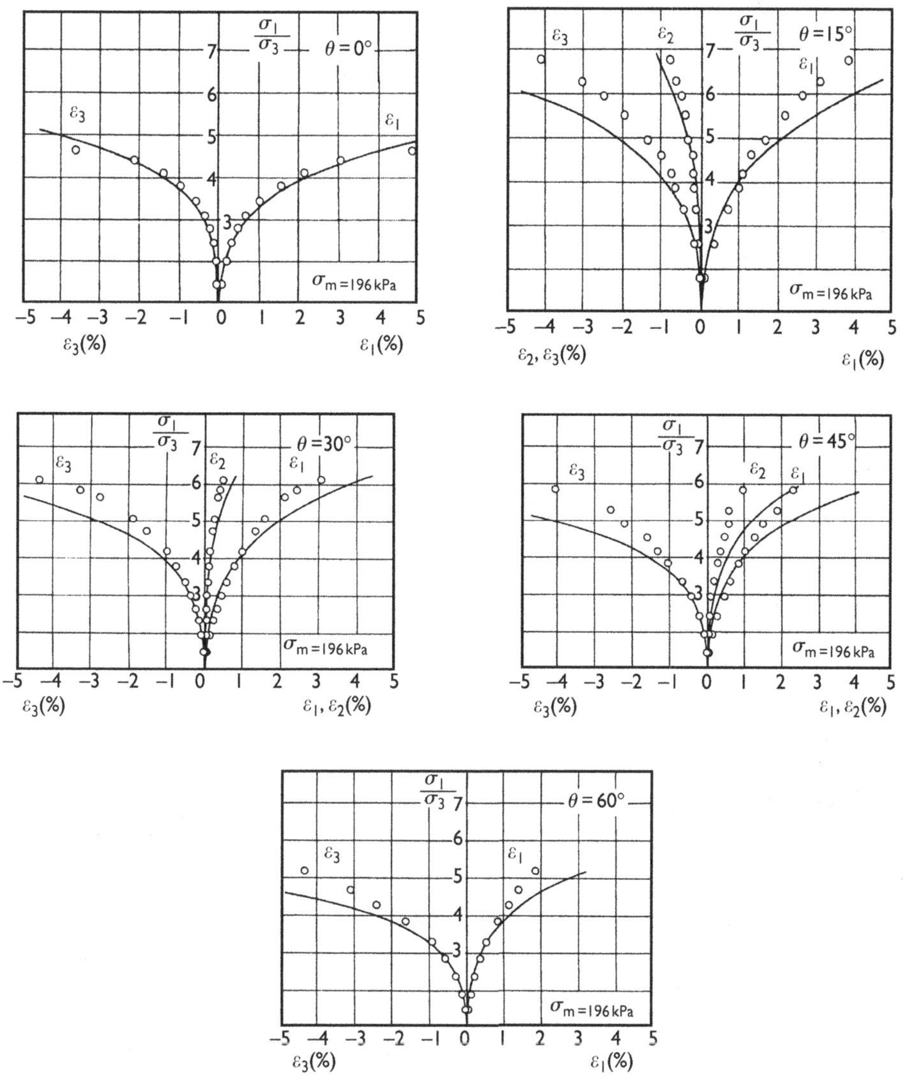

Figure 1.21 Predicted and measured relations between principal stress ratio and principal strains obtained by true triaxial tests on Toyoura sand.

between the Matsuoka-Nakai criterion and the Tresca, Mises, Mohr-Coulomb criteria are also discussed here.

Three Mohr circles for three different principal stresses are drawn in Figures 1.22 and 1.23. As is well known, there are three points (P_1, P_2, and P_3 in Fig. 1.22) on the top of the stress circles where the shear stress τ is maximized under the respective pairs of principal stresses, and three points (P_1, P_2, and P_3 in Fig. 1.23) on the tangent from the origin to the three stress circles where the shear-normal stress ratio τ/σ is maximized under the respective pairs of

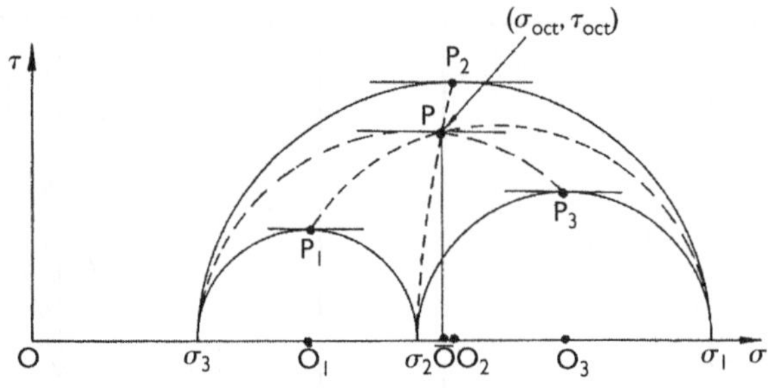

Figure 1.22 Normal and shear stresses on 45° planes and octahedral plane (Satake 1978).

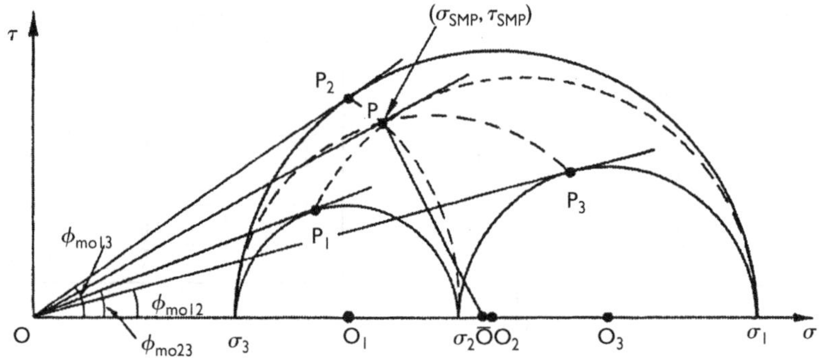

Figure 1.23 Normal and shear stresses on mobilized planes and SMP (Satake 1978).

principal stresses. The maximum shear stress τ_{max} is important for metals because the failure of metals is due to the dislocation of crystalline structure during shearing. On the other hand, the maximum shear-normal stress ratio $(\tau/\sigma)_{max}$ is important for granular materials because the failure of these materials is due to the frictional slip between particles. The planes on which τ_{max} acts are three "45° planes" shown in Figure 1.24, and the planes on which $(\tau/\sigma)_{max}$ acts are "$(45° + \phi_{moij}/2)$ planes" $(i, j = 1, 2, 3;\ i < j)$ shown in Figure 1.25(a). These "$(45° + \phi_{moij}/2)$ planes" mean the potential slip planes under the respective pairs of principal stresses, and are called "mobilized planes" (Matsuoka 1974b). For example, when ϕ_{mo13} in Figure 1.23 is equal to the internal friction angle ϕ, the plane AC in Figure 1.25(a) becomes the $(45°+\phi/2)$ plane, i.e., the slip plane (Murayama 1966). The plane whose three sides are three "45° planes" in Figure 1.24(a) is the "octahedral plane" shown in Figure 1.24(b). The plane whose three sides are three "mobilized planes" in Figure 1.25(a) is called the "Spatially Mobilized Plane (SMP)" shown in Figure 1.25(b) (Matsuoka and Nakai 1974, 1985; Matsuoka 1976). The normal stress σ_{oct} and the shear stress τ_{oct} on the octahedral plane correspond to the point P in Figure 1.22 and the normal stress σ_{SMP} and the shear stress τ_{SMP} on the SMP correspond to the point P in Figure 1.23 (Satake 1978). In Figures 1.22 and 1.23, the points O_1, O_2, O_3, and $\bar{O}$ represent $(\sigma_2 + \sigma_3)/2$, $(\sigma_3 + \sigma_1)/2$, $(\sigma_1 + \sigma_2)/2$, and $(\sigma_1 + \sigma_2 + \sigma_3)/3$ on the σ-axis.

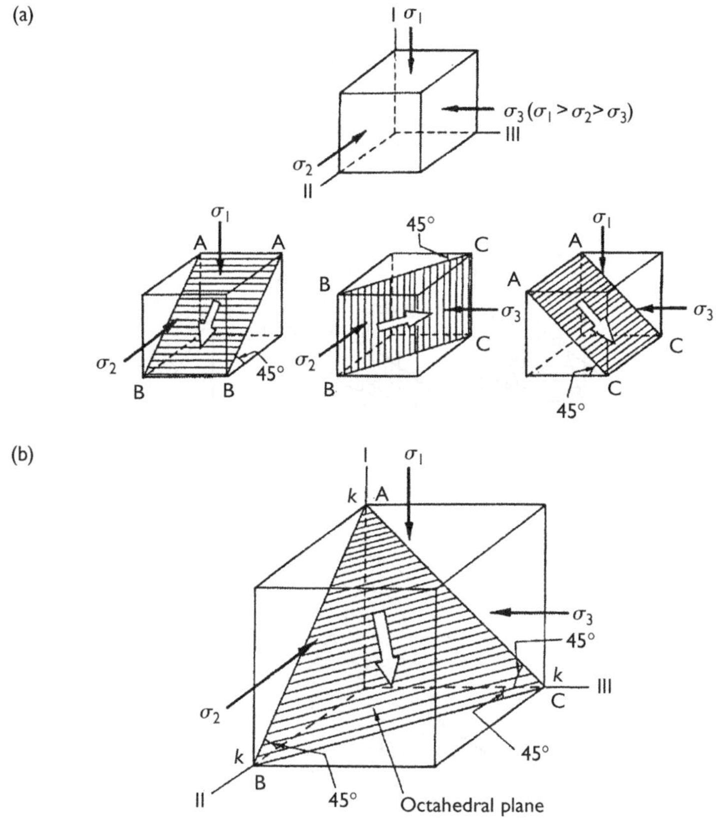

Figure 1.24 Geometrical representations of 45° planes and octahedral plane in three-dimensional space.

In Figure 1.22, the Tresca criterion which is also called the theory of the maximum shear stress is expressed as $\tau_{max} = P_2O_2 = \text{const.}$, and the Mises criterion is expressed as $\tau_{oct} = PO = \text{const.}$ In Figure 1.23, the Mohr-Coulomb criterion which is also called the theory of the maximum shear-normal stress ratio is represented by $(\tau/\sigma)_{max} = \tan\angle P_2OO_2 = \text{const.}$ (for cohesion $c = 0$), and the Matsuoka-Nakai criterion is represented by $\tau_{SMP}/\sigma_{SMP} = \tan\angle PO\bar{O} = \text{const.}$ These four types of failure criteria are expressed by the following equations.

Tresca criterion:

$$\tau_{max} = P_2O_2 = \frac{\sigma_1 - \sigma_3}{2} = \text{const.} \tag{1.19}$$

Mises criterion:

$$\tau_{oct} = \overline{PO} = \frac{2}{3}\sqrt{\left(\frac{\sigma_1 - \sigma_2}{2}\right)^2 + \left(\frac{\sigma_2 - \sigma_3}{2}\right)^2 + \left(\frac{\sigma_3 - \sigma_1}{2}\right)^2} = \text{const.} \tag{1.20}$$

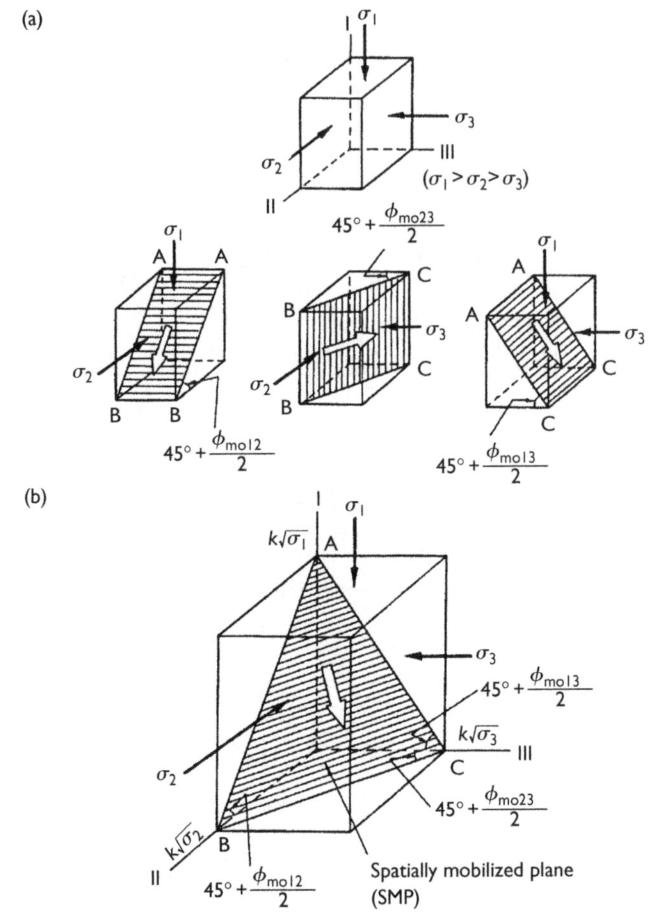

Figure 1.25 Geometrical representations of mobilized planes and SMP in three-dimensional space.

Mohr-Coulomb criterion (in the case of, $c = 0$):

$$(\tau/\sigma)_{\max} = \tan \angle \mathrm{P_2OO_2} = \frac{\sigma_1 - \sigma_3}{2\sqrt{\sigma_1\sigma_3}} = \text{const.} \tag{1.21}$$

Matsuoka-Nakai criterion (SMP criterion):

$$\begin{aligned} \tau_{\mathrm{SMP}}/\sigma_{\mathrm{SMP}} &= \tan \angle \mathrm{PO\overline{O}} \\ &= \frac{2}{3}\sqrt{\left(\frac{\sigma_1 - \sigma_2}{2\sqrt{\sigma_1\sigma_2}}\right)^2 + \left(\frac{\sigma_2 - \sigma_3}{2\sqrt{\sigma_2\sigma_3}}\right)^2 + \left(\frac{\sigma_3 - \sigma_1}{2\sqrt{\sigma_3\sigma_1}}\right)^2} = \text{const.} \end{aligned} \tag{1.22}$$

The following equivalent expression for the Matsuoka-Nakai criterion is obtained from Eq. (1.22) and Eq. (1.11)

$$I_1 I_2 / I_3 = \text{const.} \tag{1.23}$$

Where I_1, I_2, and I_3 are the first, second, and third stress invariants, which are given by Eq. (1.13). Eq. (1.23) is one of the simplest non-dimensional quantities using all of the stress invariants I_1, I_2, and I_3. The Lade-Duncan criterion (Lade and Duncan 1975) is similar to Eq. (1.23) in a form as shown by the following equation.

$$I_1^3 / I_3 = \text{const.} \tag{1.24}$$

However, the second stress invariant is not taken into account in Eq. (1.24).

It can be seen from Eqs (1.19)–(1.22) that the relation between the Matsuoka-Nakai criterion and the Mohr-Coulomb criterion is very similar to the relation between the Mises criterion and the Tresca criterion; the coefficient (2/3) on the right-hand sides of Eqs (1.20) and (1.22) are also the same. As seen from Eqs (1.19) and (1.22), the intermediate principal stress σ_2 is not taken into account in the Tresca and the Mohr-Coulomb criteria, but are taken into account in the Mises and Matsuoka-Nakai criteria. It is clear that the Mohr-Coulomb criterion is a two-dimensional frictional law determined by two principal stresses, but the Matsuoka-Nakai criterion is a three-dimensional frictional law determined by three principal stresses. Therefore, the Matsuoka-Nakai criterion can be considered as a generalization of the Mohr-Coulomb criterion into the three dimensions under three principal stresses, in the same way as the Mises criterion generalizes the Tresca criterion.

Figure 1.26 shows the relationships between these failure criteria described on the π-plane. Just as the Mises circle circumscribes the regular Tresca hexagon, the Matsuoka-Nakai criterion is a smooth convex curve circumscribing the irregular hexagon of the Mohr-Coulomb criterion. The shapes of the Tresca and Mises criteria, and the Mohr-Coulomb and Matsuoka-Nakai criteria in the principal stress space are shown in Figure 1.27. It is seen from Figure 1.27(b) that the Mohr-Coulomb and Matsuoka-Nakai criteria become zero at the origin. The reason is that the shear strength becomes zero under the zero confining pressure ($\sigma_1 = \sigma_2 = \sigma_3 = 0$) when the cohesion of the soil is zero.

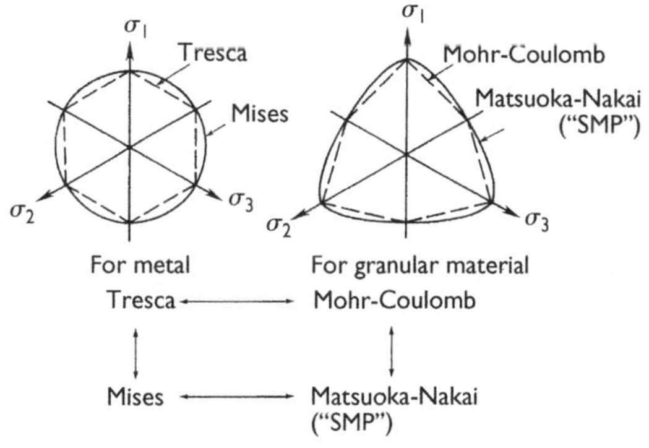

Figure 1.26 Mutual relationships between the Tresca, Mises, Mohr-Coulomb, and Matsuoka-Nakai (SMP) failure criteria described in the π-plane.

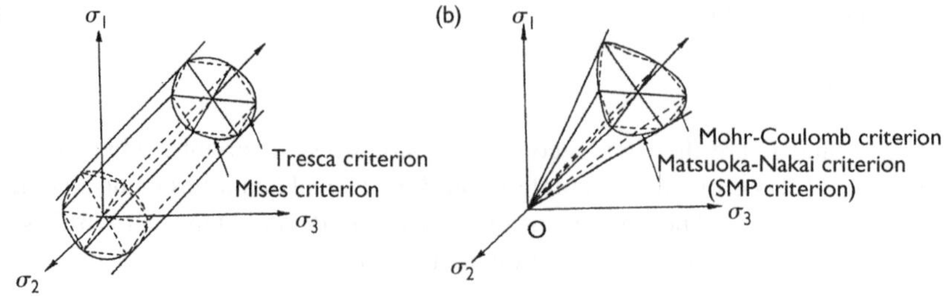

Figure 1.27 Shapes of the Tresca, Mises, Mohr-Coulomb, and Matsuoka-Nakai (SMP) failure criteria in principal stress space.

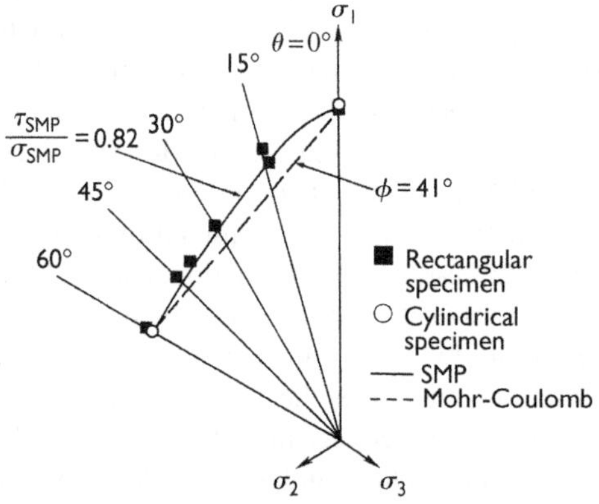

Figure 1.28 Comparison of the Matsuoka-Nakai (SMP) and Mohr-Coulomb criteria with stress states at failure in the π-plane obtained by triaxial compression, triaxial extension and true triaxial tests on Toyoura sand.

Figure 1.28 shows the Mohr-Coulomb and Matsuoka-Nakai failure criteria in the π-plane, compared with the failure stress states obtained by the triaxial compression, triaxial extension, and true triaxial tests on Toyoura sand (Matsuoka and Nakai 1982; Nakai and Matsuoka 1983). It can be seen from Figure 1.28 that the measured values agree well with the Matsuoka-Nakai failure criterion represented by the solid curve. It is clear that the Matsuoka-Nakai (SMP) failure criterion is capable of predicting the strength of soils under three-dimensional stress states.

1.5 Failure criterion for cohesive-frictional materials

In this section, we will discuss the failure criterion for a wide range of engineering materials from frictional materials such as sands without bond between particles to cohesive materials such as metals with a strong bond due to their crystalline structure. The SMP for frictional materials is extended to a plane called the "Extended Spatially Mobilized Plane (Extended SMP)" for cohesive-frictional materials, by introducing a parameter called "bonding stress."

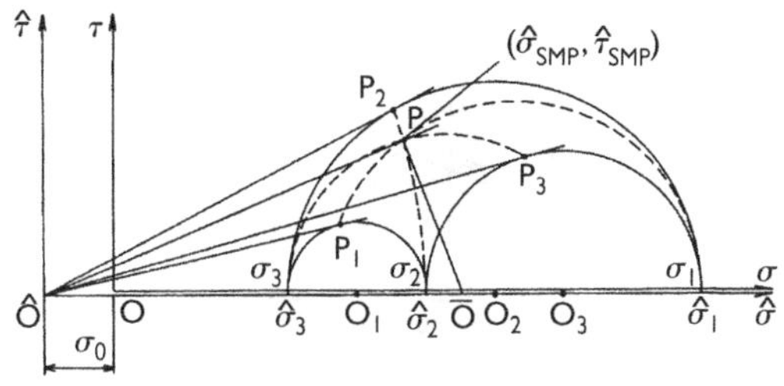

Figure 1.29 Normal and shear stresses on mobilized planes and Extended SMP for cohesive-frictional materials.

In order to extend the SMP failure criterion for frictional materials to that for cohesive-frictional materials, we draw three tangent lines from a point Ô on the negative side of the normal stress axis to the three Mohr's circles, as shown in Figure 1.29. The absolute value of the normal stress at Ô is called the bonding stress σ_0 and can be expressed as

$$\sigma_0 = c \cdot \cot \phi \tag{1.25}$$

where c is the cohesion and ϕ is the internal friction angle. For cohesive materials such as metals with $\phi = 0$ and $0 < c < \infty$, we have $\sigma_0 \to \infty$ from the definition of σ_0 in Eq. (1.25). On the other hand, for frictional materials such as granular materials with $c = 0$ and $0 < \phi < \infty$, we have $\sigma_0 = 0$.

If the point Ô is considered as the origin of a new normal–shear stress coordinate system (Fig. 1.29), we can get a failure criterion for cohesive-frictional materials by following the SMP criterion. Using the new coordinate system, Matsuoka *et al.* (1990) introduced a translated principal stress $\hat{\sigma}$ and the translated stress invariants $\hat{I}_1$, $\hat{I}_2$, and $\hat{I}_3$:

$$\hat{\sigma}_i = \sigma_i + \sigma_0 \quad (i = 1, 2, 3) \tag{1.26}$$

$$\left.\begin{aligned} \hat{I}_1 &= \hat{\sigma}_1 + \hat{\sigma}_2 + \hat{\sigma}_3 \\ \hat{I}_2 &= \hat{\sigma}_1\hat{\sigma}_2 + \hat{\sigma}_2\hat{\sigma}_3 + \hat{\sigma}_3\hat{\sigma}_1 \\ \hat{I}_3 &= \hat{\sigma}_1\hat{\sigma}_2\hat{\sigma}_3 \end{aligned}\right\} \tag{1.27}$$

Figure 1.30 shows the plane on which $\hat{\sigma}_{SMP}$ and $\hat{\tau}_{SMP}$ act. This plane is named the Extended SMP. The direction cosines $\hat{a}_i$ of the normal to the "Extended SMP" are expressed as

$$\hat{a}_i = \sqrt{\hat{I}_3/(\hat{\sigma}_i\hat{I}_2)} \quad (i = 1, 2, 3) \tag{1.28}$$

It should be noted from the given equation that (1) when $\sigma_0 = 0$, $\hat{a}_i = \sqrt{I_3/(\sigma_i I_2)} = a_i$, the direction cosines of the normal to the SMP recover and (2) when $\sigma_0 \to \infty$, $\hat{a}_i = 1/\sqrt{3}$, the direction cosines of the normal to the octahedral plane recover. The normal stress $\hat{\sigma}_{SMP}$ and the shear stress $\hat{\tau}_{SMP}$ on the "Extended SMP" under three translated principal stresses $\hat{\sigma}_i$

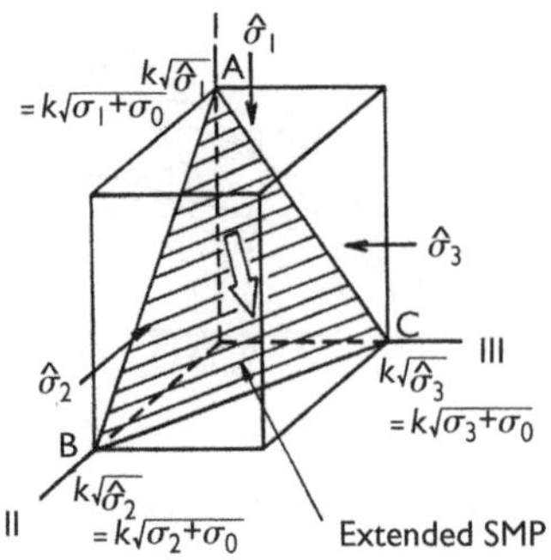

Figure 1.30 Extended SMP in three-dimensional space under translated principal stresses.

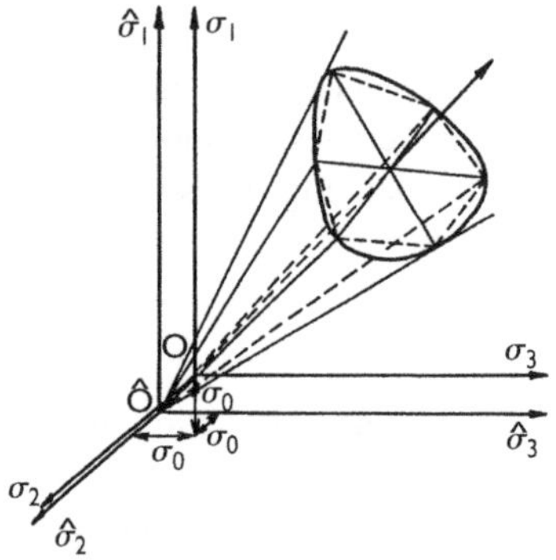

Figure 1.31 Schematic diagrams of Extended SMP failure criterion and general Mohr-Coulomb failure criterion in principal stress space.

are represented as follows (Matsuoka *et al.* 1990):

$$\hat{\sigma}_{\mathrm{SMP}} = \hat{\sigma}_1\hat{a}_1^2 + \hat{\sigma}_2\hat{a}_2^2 + \hat{\sigma}_3\hat{a}_3^2 \tag{1.29}$$

$$\hat{\tau}_{\mathrm{SMP}} = \sqrt{(\hat{\sigma}_1 - \hat{\sigma}_2)^2\hat{a}_1^2\hat{a}_2^2 + (\hat{\sigma}_2 - \hat{\sigma}_3)^2\hat{a}_2^2\hat{a}_3^2 + (\hat{\sigma}_3 - \hat{\sigma}_1)^2\hat{a}_3^2\hat{a}_1^2} \tag{1.30}$$

When the shear–normal stress ratio $\hat{\tau}_{\mathrm{SMP}}/\hat{\sigma}_{\mathrm{SMP}}$ on the "Extended SMP" reaches a critical value, the material is considered to fail. A new failure criterion can thus be derived from Eqs (1.25) to (1.30) and is expressed as follows:

$$\begin{aligned} \frac{\hat{\tau}_{\mathrm{SMP}}}{\hat{\sigma}_{\mathrm{SMP}}} &= \sqrt{\frac{\hat{I}_1\hat{I}_2 - 9\hat{I}_3}{9\hat{I}_3}} \\ &= \frac{2}{3}\sqrt{\frac{(\sigma_1-\sigma_2)^2}{4(\sigma_1+\sigma_0)(\sigma_2+\sigma_0)} + \frac{(\sigma_2-\sigma_3)^2}{4(\sigma_2+\sigma_0)(\sigma_3+\sigma_0)} + \frac{(\sigma_3-\sigma_1)^2}{4(\sigma_3+\sigma_0)(\sigma_1+\sigma_0)}} \\ &= \text{const.} \end{aligned} \tag{1.31}$$

The Extended SMP failure criterion defined by Eq. (1.31) recovers to the Matsuoka-Nakai criterion with $c = 0$ or $\sigma_0 = 0$ and the Mises criterion with $\phi = 0$ or $\sigma_0 \to \infty$. Figure 1.31

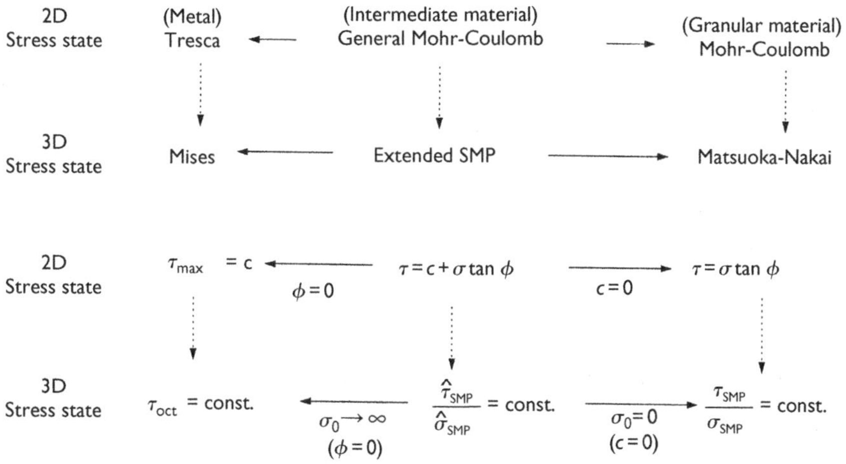

Figure 1.32 Mutual relationships between the Tresca, Mises, Mohr-Coulomb, generalized Mohr-Coulomb, Matsuoka-Nakai (SMP), and Extended SMP failure criteria.

shows the shape of the Extended SMP failure criterion in the three-dimensional stress space. Figure 1.32 shows the relationships between the Tresca, Mises, Mohr-Coulomb, generalized Mohr-Coulomb, Matsuoka-Nakai, and Extended SMP failure criteria. It can be seen that, just as the Mohr-Coulomb criterion for frictional materials with $c = 0$ and the Tresca criterion for cohesive materials with $\phi = 0$ correspond to the two extremes of the generalized Mohr-Coulomb criterion in two-dimensional (2D) stresses, the Matsuoka-Nakai criterion for frictional materials ($\sigma_0 = 0$ and $c = 0$) and the Mises criterion for cohesive materials ($\phi = 0$ and $\sigma_0 \to \infty$) correspond to two extremes of the Extended SMP criterion in three-dimensional (3D) stresses. It can also be concluded from Figure 1.32 that, just as the Matsuoka-Nakai criterion is to the Mohr-Coulomb criterion under three principal stresses and the Mises criterion to the Tresca criterion (Matsuoka and Nakai 1985), the Extended SMP has the concept of averaging the generalized Mohr-Coulomb criterion under three principal stresses. With the help of the criterion based on the Extended SMP, therefore, it is possible to explain the failure of cohesive-frictional materials, which cover a wide range of engineering materials from frictional materials such as soils to cohesive materials such as metals. To investigate this possibility, triaxial compression, triaxial extension, and true triaxial tests have been carried out on cemented sands (Matsuoka and Sun 1995), which can be considered as an intermediate material with bond and cohesion.

Figure 1.33 shows a comparison between the Extended SMP failure criterion in the π-plane and the failure stress states obtained by true triaxial, triaxial compression, and triaxial extension tests on a cemented sand specimen (Matsuoka and Sun 1995). The component ratios by weight for the cemented sand are Toyoura sand : cement : water = 15 : 1 : 3. The curing period for the cemented sand specimens is about three months. The measured stress points are marked by □ for the true triaxial tests, ○ for the triaxial compression and extension tests. The solid curve is predicted by the Extended SMP failure criterion. It can be seen that the predictions agree well with the test results.

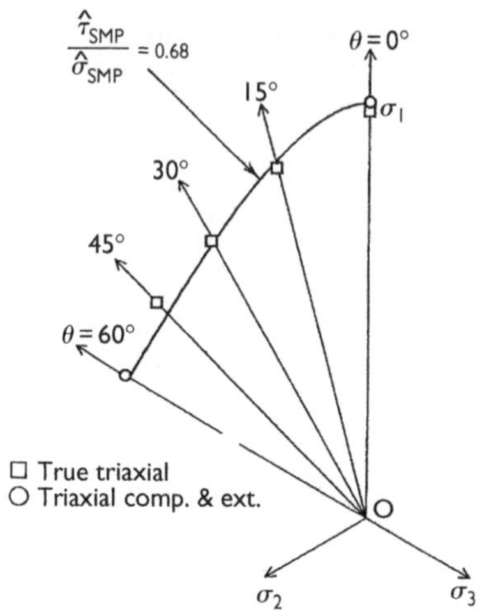

Figure 1.33 Comparison of the Extended SMP criterion with stress states at failure in the π-plane obtained by triaxial compression, triaxial extension, and true triaxial tests on a cemented sand.

References

Lade, P.V. and Duncan, J.M. 1975. Elasto-plastic stress-strain theory for cohesionless soil. *Journal of Geotechnical Engineering ASCE*, 101(10): 1037–1053.

Matsuoka, H. 1974a. A microscopic study on shear mechanism of granular materials. *Soils and Foundations*, 14(1): 29–43.

Matsuoka, H. 1974b. Stress-strain relationship of sands based on the mobilized plane. *Soils and Foundations*, 14(2): 47–61.

Matsuoka, H. 1976. On the significance of the spatial mobilized plane. *Soils and Foundations*, 16(1): 91–100.

Matsuoka, H. 1983. Deformation and strength of granular materials based on the theory of "compounded mobilized plane (CMP)" and "spatial mobilized plane"(SMP). In *Advances in the Mechanics and the Flow of Granular Materials (II)*. Trans. Tech. Publication, Clausthal-Zellerfeld, Germany, pp. 813–836.

Matsuoka, H. and Nakai, T. 1974. Stress-deformation and strength characteristics of soil under three different principal stresses. *Proceedings of JSCE*, 232: 59–74.

Matsuoka, H. and Nakai, T. 1985. Relationship among Tresca, Mises, Mohr-Coulomb and Matsuoka-Nakai failure criteria. *Soils and Foundations*, 25(4): 123–128.

Matsuoka, H. and Sun, D.A. 1995. Extension of spatially mobilized plane (SMP) to frictional and cohesive materials and its application to cemented sands. *Soils and Foundations*, 35(4): 63–72.

Matsuoka, H., Hoshikawa, T., and Ueno, K. 1990. A general failure criterion and stress-strain relation for granular materials to metals. *Soils and Foundations*, 30(2): 119–127.

Murayama, S. 1966. A theoretical consideration on behaviour of sand. Proceedings of the IUTAM Symposium on Rheology and Soil Mechanics, Grenoble, pp. 146–159.

Nakai, T. and Matsuoka, H. 1982. Deformation of soils in three-dimensional stresses. Proceedings of the IUTAM Symposium on Deformation and Failure of Granular materials, Delft, pp. 275–285.

Nakai, T. and Matsuoka, H. 1983. Shear behavior of sand and clay under three-dimensional stress condition. *Soils and Foundations*, 23(2): 26–42.

Rowe, P.W. 1962. The stress–dilatancy relation for static equilibrium of an assembly of particle in contact. *Proceedings of the Royal Society*, A. 269: 500–527.

Satake, M. 1978. Comment by Satake, Proceedings of the US–Japan Seminar on Continuum Mechanical and Statistical Approaches in the Mechanics of granular materials, Gakujyutsu Bunken Fukyu-kai, p. 154.

Chapter 2

Introduction to Cam-clay model

2.1 Introduction

So far, we have discussed the concept of SMP based on micro-observation of particle movement in soils. In this and subsequent chapters, we will discuss constitutive models for soils using a macro-approach, i.e., the plasticity theory.

The Cam clay is not a real soil, but an ideal soil proposed by Roscoe *et al.* (1963). The Cam-clay model is an elastoplastic constitutive model for normally consolidated clay, and is one of the early constitutive models that adopt the metal plasticity theory. It has had a tremendous influence on subsequent studies of soil constitutive modeling. Now, the Cam-clay model is considered a classic model in the field of the constitutive studies of geomaterials. Although there are many reviews (e.g. Tatsuoka 1978; Ohta 1993) and text books (e.g. Schofield and Wroth 1968; Wood 1990) about the Cam-clay model, its concepts are not easy to understand for beginners.

Here, we try to explain the Cam-clay model as easily as possible, in particular from the point of view of how it can predict strain. To do this, let us first review some fundamental concepts used in the plasticity theory: (a) plastic potential, (b) yield function, and (c) strain-hardening rule.

The plastic potential function is used to determine the direction of the plastic strain increment, i.e., the plastic strain increment is normal to the plastic potential. The yield function is used to distinguish the elastic region from the plastic region in the principal stress space. The strain-hardening rule is used as an evolution law for the yield surfaces, which are a family of the surfaces expressed by the yield function in the principal stress space, i.e., a rule to indicate the hardening degree.

In the Cam-clay model, first, the relation between the stress ratio and strain increment ratio is derived from the energy dissipation equation. This relation seems to be a true stress–strain relation for soils. An alternative method to obtain the relation is from the balance of contact force between particles on the mobilized plane (e.g. Newland and Allely 1957; Rowe 1962). Although it may be difficult to know where the soil particles move under a given stress ratio, it is possible, to some extent, to know the average of the contact angles of sliding particles, i.e., the strain increment ratio. Since the elastic component is very small, the total strain increment ratio could be considered to be the plastic strain increment ratio, i.e., the direction of the plastic strain increment. When the direction is known, we can obtain a surface normal to the plastic strain increment vectors in principal stress space using mathematical methods. This surface is then the so-called plastic potential surface and its expression is the plastic potential function.

If the yield function is assumed to be the same as the plastic potential function, we then have a so-called associated flow rule. The yield function can be depicted in the principal stress space to form a surface, which is called the yield surface. When the stress state is inside the yield surface, the stress change yields elastic deformation only. When the stress state is outside the yield surface, the stress change yields elastic and plastic deformations.

In the discussion here, we use only the relation between the stress ratio and the strain increment ratio, and hence cannot determine the strain. If, for example, one of two strains is unknown, the other strain cannot be determined even if the strain increment ratio is known. Hence, we need to know the strain-hardening rule. Figure 2.1 shows the water content contours of normally consolidated clay, obtained from drained and undrained triaxial compression and extension tests (Henkel 1960). In the case of full saturation, the water content contour is equivalent to the volumetric strain contour. Accounting for the similarity between the contour and the yield locus, the Cam-clay model adopts the volumetric strain as the hardening parameter.

It is convenient to determine the values of the hardening parameter (i.e., the volumetric strain) using the stress–strain relation along a simplest stress path, i.e., the isotropic stress path. In the Cam-clay model, a linear e-log p relation is used in the isotropic stress, and then the volumetric strain in any stress state can be calculated using the yield function, and finally from the stress ratio versus strain increment ratio relation the plastic shear strain can be calculated using the known volumetric strain. When the volumetric and shear strains are known, all plastic strains can be calculated under the assumption of the coaxiality of the axes of stress and plastic strain increments.

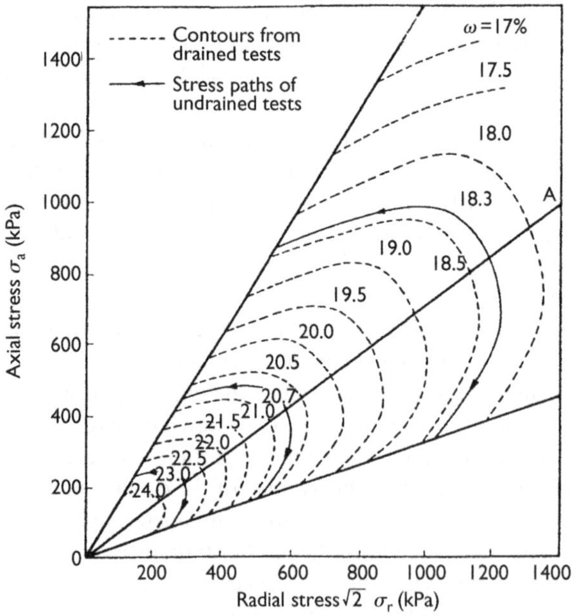

Figure 2.1 Water content contours from drained triaxial tests and stress paths in undrained triaxial tests on a normally consolidated clay.

Therefore, there are two key points in the model: (a) the relation between the stress ratio and the strain increment ratio, and (b) the determination of the volumetric strain in the isotropic stress states using a linear e-log p relation and its extension to the general stress states on the basis of Henkel's water content contours. In the following sections, we will introduce the Cam-clay model in detail with special emphasis on these two key points.

2.2 Original Cam-clay model

The Cam-clay model was developed by Roscoe *et al.* at Cambridge University in the 1960s; it is a conceptual elastoplastic model that can represent the mechanical behavior of normally consolidated clays.

In the elastoplastic constitutive model, the total strain increments are decomposed into elastic and plastic strain increments, i.e.,

$$d\varepsilon_{ij} = d\varepsilon_{ij}^{\mathrm{e}} + d\varepsilon_{ij}^{\mathrm{p}} \tag{2.1}$$

where the elastic strain increment $d\varepsilon_{ij}^{\mathrm{e}}$ is calculated by Hooke's law for the elastic behavior of soil. The plastic strain increment $d\varepsilon_{ij}^{\mathrm{p}}$ is determined based on the plasticity theory and the deformation and strength characteristics of the soil. In the plasticity theory, the following key concepts are necessary to determine the plastic strain increments:

(a) *Principal direction of the plastic strain increment*: Most of the elastoplastic models adopt the assumption that the principal directions of stresses and plastic strain increments are coaxial.
(b) *Determinations of the plastic potential and yield function*: The plastic potential defines the direction of the plastic strain increment. The yield function specifies whether plastic strain increments occur when subjected to a new loading increment.
(c) *Determination of strain-hardening rule*: The strain-hardening rule affects the magnitude of the plastic strain increment. Therefore, the Cam-clay model covers these concepts for the mechanical behavior of normally consolidated clays. In the following section, we will discuss it in turn.

2.2.1 Principal direction of plastic strain increment and adopted stress and strain variables

Similar to other elastoplastic models, the Cam-clay model assumes the coaxiality between the principal directions of stress σ_{ij} and plastic strain increment $d\varepsilon_{ij}^{\mathrm{p}}$ (see Fig. 2.2). Since the direction cosines of the normal to the octahedral plane are $(1/\sqrt{3}, 1/\sqrt{3}, 1/\sqrt{3})$, as shown in Figure 2.2(a), the lengths of vectors $\overrightarrow{\mathrm{OP}}$ and $\overrightarrow{\mathrm{PS}}$ can be written as

$$\left|\overrightarrow{\mathrm{OP}}\right| = \sigma_1\frac{1}{\sqrt{3}} + \sigma_2\frac{1}{\sqrt{3}} + \sigma_3\frac{1}{\sqrt{3}} = \sqrt{3}\frac{\sigma_1+\sigma_2+\sigma_3}{3} = \sqrt{3}p \tag{2.2}$$

$$\begin{aligned}\left|\overrightarrow{\mathrm{PS}}\right| &= \sqrt{\left|\overrightarrow{\mathrm{OS}}\right|^2 - \left|\overrightarrow{\mathrm{OP}}\right|^2} = \sqrt{\sigma_1^2+\sigma_2^2+\sigma_3^2 - \{(1/\sqrt{3})(\sigma_1+\sigma_2+\sigma_3)\}^2} \\ &= \frac{1}{\sqrt{3}}\sqrt{(\sigma_1-\sigma_2)^2+(\sigma_2-\sigma_3)^2+(\sigma_3-\sigma_1)^2} = \sqrt{\frac{2}{3}}q\end{aligned} \tag{2.3}$$

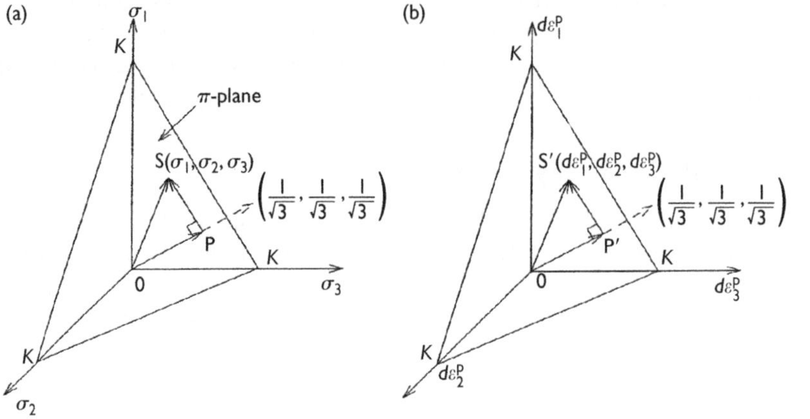

Figure 2.2 (a) Principal stress space and (b) principal plastic strain increment space.

where,

$$p = \frac{\sigma_1 + \sigma_2 + \sigma_3}{3} \tag{2.4}$$

$$q = \frac{1}{\sqrt{2}}\sqrt{(\sigma_1 - \sigma_2)^2 + (\sigma_2 - \sigma_3)^2 + (\sigma_3 - \sigma_1)^2} \tag{2.5}$$

In addition, the mean stress p or σ_m and the octahedral normal stress σ_{oct} are identical, i.e., $p = \sigma_{oct} = \sigma_m$; and the deviator stress q is $3/\sqrt{2}$ times of the octahedral shear stress τ_{oct}, i.e., $q = 3/\sqrt{2}\tau_{oct}$.

On the other hand, as shown in Figure 2.2(b), the lengths of vectors $\overrightarrow{OP'}$ and $\overrightarrow{P'S'}$ are as follows:

$$\left|\overrightarrow{OP'}\right| = d\varepsilon_1^p \frac{1}{\sqrt{3}} + d\varepsilon_2^p \frac{1}{\sqrt{3}} + d\varepsilon_3^p \frac{1}{\sqrt{3}} = \frac{1}{\sqrt{3}}(d\varepsilon_1^p + d\varepsilon_2^p + d\varepsilon_3^p) = \frac{1}{\sqrt{3}} d\varepsilon_v^p \tag{2.6}$$

$$\left|\overrightarrow{P'S'}\right| = \sqrt{\left|\overrightarrow{OS'}\right|^2 - \left|\overrightarrow{OP'}\right|^2} = \sqrt{d\varepsilon_1^{p2} + d\varepsilon_2^{p2} + d\varepsilon_3^{p2} - \left\{\frac{1}{\sqrt{3}}(d\varepsilon_1^p + d\varepsilon_2^p + d\varepsilon_3^p)\right\}^2}$$

$$= \frac{1}{\sqrt{3}}\sqrt{(d\varepsilon_1^p - d\varepsilon_2^p)^2 + (d\varepsilon_2^p - d\varepsilon_3^p)^2 + (d\varepsilon_3^p - d\varepsilon_1^p)^2} = \sqrt{\frac{3}{2}} d\varepsilon_d^p \tag{2.7}$$

where,

$$d\varepsilon_v^p = d\varepsilon_1^p + d\varepsilon_2^p + d\varepsilon_3^p \tag{2.8}$$

$$d\varepsilon_d^p = \frac{\sqrt{2}}{3}\sqrt{(d\varepsilon_1^p - d\varepsilon_2^p)^2 + (d\varepsilon_2^p - d\varepsilon_3^p)^2 + (d\varepsilon_3^p - d\varepsilon_1^p)^2} \tag{2.9}$$

are the plastic volumetric strain increment and the plastic deviatoric strain increment respectively. Note that $d\varepsilon_d^p$ is $1/\sqrt{2}$ times of the plastic shear strain increment $d\gamma_{oct}^p$ on the octahedral plane, i.e., $d\varepsilon_d^p = 1/\sqrt{2} d\gamma_{oct}^p$.

The stress variables adopted in the Cam-clay model are p and q, and the plastic strain increments are $d\varepsilon_v^p$ and $d\varepsilon_d^p$.

2.2.2 Determination of plastic potential and yield functions

Because the coaxiality between the stress σ_{ij} and the plastic strain increment $d\varepsilon_{ij}^p$ is assumed, we can superimpose the principal axes for the stress space and for the plastic strain increment space, as shown in Figure 2.2.

In order to determine the plastic potential function, we use the normality condition, i.e., there is a family of surfaces normal to the plastic strain increment vector in the stress space. This normality condition, together with Eqs (2.2), (2.3), (2.6), and (2.7) leads to

$$d\sigma_1\, d\varepsilon_1^p + d\sigma_2\, d\varepsilon_2^p + d\sigma_3\, d\varepsilon_3^p = d\left|\overrightarrow{OP}\right| \cdot \left|\overrightarrow{OP'}\right| + d\left|\overrightarrow{PS}\right| \cdot \left|\overrightarrow{P'S'}\right|$$

$$= \sqrt{3}dp \cdot \frac{1}{\sqrt{3}} d\varepsilon_v^p + \sqrt{\frac{2}{3}} dq \cdot \sqrt{\frac{3}{2}} d\varepsilon_d^p = dp \cdot d\varepsilon_v^p + dq \cdot d\varepsilon_d^p = 0 \tag{2.10}$$

Eq. (2.10) can be rewritten as

$$\frac{dq}{dp} \times \frac{d\varepsilon_d^p}{d\varepsilon_v^p} = -1 \quad \text{(normality condition)} \tag{2.11}$$

As shown in Figure 2.3, when we superimpose the p-axis to the $d\varepsilon_v^p$-axis, the q-axis to the $d\varepsilon_d^p$-axis, Eq. (2.10) actually defines the surfaces that are normal to the plastic strain increment vectors. This surface is the plastic potential ($g = 0$).

To determine the plastic potential surfaces, we need the relation between the direction of the plastic strain increment vector and the stresses p and q. In order to obtain this relation, the energy dissipation equation was used in the Cam-clay model.

The increment of external work per unit volume can be expressed by

$$dW^p = \sigma_1 d\varepsilon_1^p + \sigma_2 d\varepsilon_2^p + \sigma_3 d\varepsilon_3^p = pd\varepsilon_v^p + qd\varepsilon_d^p \tag{2.12}$$

The second equal sign in Eq. (2.12) can be understood by considering Eq. (2.10). In the Cam-clay model, $q_f = Mp$ and $d\varepsilon_v^p = 0$ are adopted at failure. Substituting this condition into Eq. (2.12) leads to

$$dW^p = pd\varepsilon_v^p + qd\varepsilon_d^p = q_f d\varepsilon_d^p = Mpd\varepsilon_d^p \tag{2.13}$$

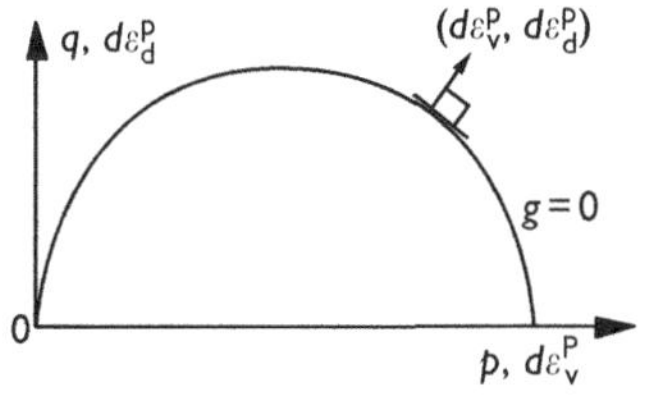

Figure 2.3 Plastic potential and strain increment vector.

This equation seems perfect, but there is a strong assumption in it. The first equal sign in Eq. (2.13) is always tenable, but the second and third equal signs hold only at failure. That is to say, the failure condition was applied to the full shear process from the initial stage to failure; therefore, this is a strong assumption.

Rearranging Eq. (2.13) gives

$$\frac{q}{p} = M - \frac{d\varepsilon_{\mathrm{v}}^{\mathrm{p}}}{d\varepsilon_{\mathrm{d}}^{\mathrm{p}}} \tag{2.14}$$

This equation represents the relation between the stress ratio and the plastic strain increment ratio in the Cam-clay model, and also defines the relation between the direction of the plastic strain increment vector and the stress ratio (q/p). As mentioned earlier, from author's point of view, Eq. (2.14) is the first key point in the Cam-clay model and is depicted as shown in Figure 2.4.

Combining the relation between the direction of plastic strain increment vector and stress ratio (Eq. (2.14)) with the normality condition (Eq. (2.11)), we can obtain the plastic potential function in the p–q plane. Substituting Eq. (2.11) into Eq. (2.14) gives

$$\frac{dq}{dp} + M - \frac{q}{p} = 0 \tag{2.15}$$

Exercise 2.1 Solve the ordinary differential equation (2.15).

Answer: Letting $q/p = x$ gives

$$q = px$$

$$\frac{dq}{dp} = \frac{d(px)}{dp} = x + p\frac{dx}{dp}$$

So, Eq. (2.15) can be written as

$$x + p\frac{dx}{dp} + M - x = 0$$

$$\frac{dx}{dp} + \frac{M}{p} = 0$$

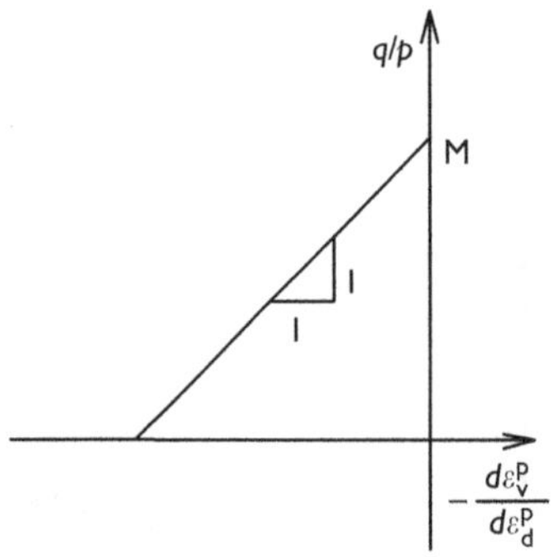

Figure 2.4 Stress ratio versus strain increment ratio relation adopted in Cam-clay model.

$$dx + M\frac{dp}{p} = 0$$

$$\int dx + M\int \frac{dp}{p} = C$$

$$x + M \ln p = C$$

$$\therefore M \ln p + \frac{q}{p} - C = 0 \quad (C\text{: an integral constant})$$

This equation shows a family of the curves orthogonal to the plastic strain increment vectors whose direction is expressed by Eq. (2.14), and this family of curves are of course the plastic potential function we are after.

$$g = M \ln p + \frac{q}{p} - C = 0 \tag{2.16}$$

where C is an integral constant, and ln means the natural logarithm.

Figure 2.5 shows the relation between the plastic strain increment vectors indicated by Eq. (2.14) and the plastic potential ($g = 0$, dotted line). From Eq. (2.14), we know that $d\varepsilon_d^p/d\varepsilon_v^p = 1/M$ when $q/p = 0$ and $d\varepsilon_v^p = 0$ when $q/p = M$, which is called the Critical State Line (CSL).

In the Cam-clay model, the associated flow rule is adopted, i.e., the yield function f is equal to the plastic potential function g. Hence, from Eq. (2.16) the yield function f is expressed as

$$f = M \ln p + \frac{q}{p} - C = 0 \tag{2.17}$$

Letting $p = p_x$ when $q = 0$ results in $C = M \ln p_x$, so Eq. (2.17) can be rewritten as

$$f = M \ln p + \frac{q}{p} - M \ln p_x = 0 \tag{2.18}$$

Figure 2.6 shows the shape of the curve expressed by the yield function $f = 0$ (Eq. 2.18), which is called the yield locus. If the current stress state is at the yield locus, only elastic strain occurs when the new stress state is inside the yield locus; elastoplastic strain occurs

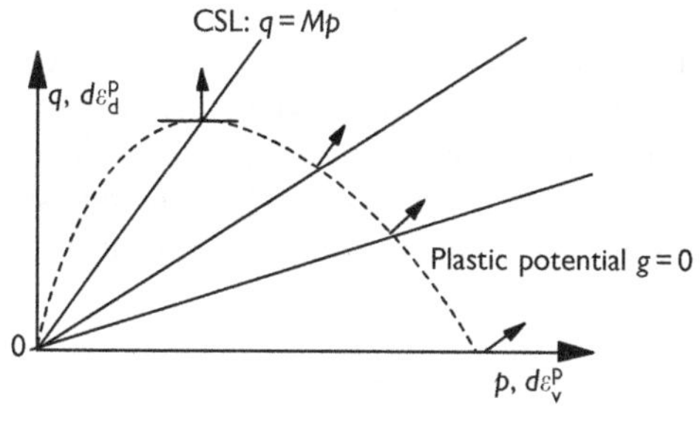

Figure 2.5 Plastic potential and plastic strain increment vectors in Cam-clay model.

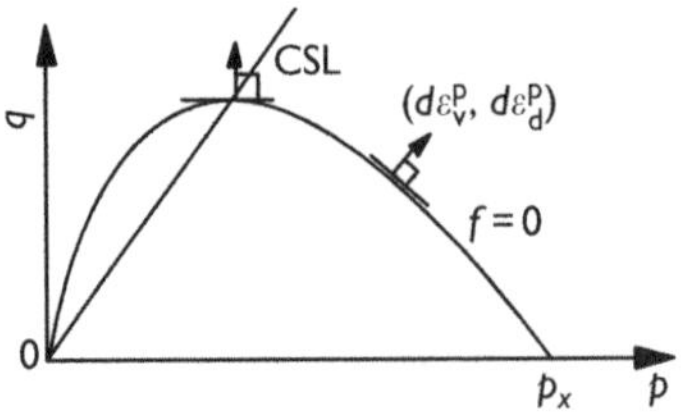

Figure 2.6 Yield locus ($f = 0$) used in Cam-clay model.

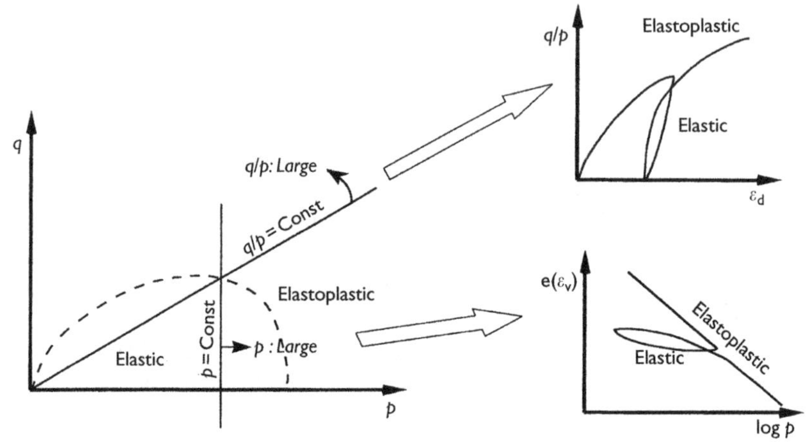

Figure 2.7 The mechanical meaning of yield locus.

when the new stress state is outside the current yield locus. When p_x becomes large the yield locus expands similarly.

Here, let us discuss the mechanical meaning of the yield locus.

Figure 2.7 depicts the yield loci for q/p = constant during shearing and for p = constant during consolidation. As is well known, when the stress ratio q/p (or τ/σ) becomes large, a large strain will occur since soil is a frictional material, which can be considered to be largely plastic; when the stress ratio becomes small, only a small strain occurs, which can be considered to be mainly elastic. On the other hand, when the mean stress p becomes large, a large consolidated strain will occur, which is largely plastic; and when the mean stress p becomes small, only a small strain occurs, which is mainly elastic. So, q/p = constant and p = constant can be considered to be respective yield loci. These two yield loci results are expressed by one yield locus as the dotted curve in Figure 2.7. Therefore, it can be concluded that the mechanical meaning of the yield locus such as the one shown in Figure 2.6 is to describe the plastic strains when q/p and/or p becomes large using one yield locus.

2.2.3 *Determination of strain-hardening rule*

When the plastic strain occurs, knowing the plastic strain increment ratio is not sufficient for determining the magnitude of plastic strain. It is necessary to introduce a so-called

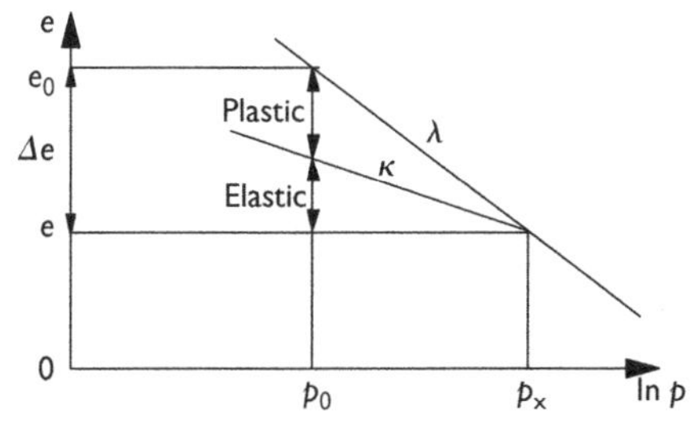

Figure 2.8 Results of isotropic compression and swelling tests.

strain-hardening rule. In the Cam-clay model, the results of isotropic compression tests are used to derive the strain-hardening rule. The isotropic compression test is a simple test that measures the stress–strain relation of a soil under isotropic stress state. As is well known, the test results are usually plotted in the e-log p diagram and the virgin-loading curve and unloading-reloading curve are assumed to be straight lines, with the compression index C_c and the swelling index C_s denoting the slopes of the two lines, respectively. In the e-ln p diagram (Fig. 2.8), the slopes of the two lines are usually denoted by λ and κ for the isotropic virgin-loading condition and unloading-reloading condition. The e-ln p relation under isotropic unloading and reloading are considered to be identical and elastic. The relations between the compression indexes C_c and λ, and the swelling indexes C_s and κ are $\lambda = 0.434C_c$ and $\kappa = 0.434C_s$.

As shown in Figure 2.8, the following relations can be obtained.

$$\Delta e = e - e_0 = -\lambda \ln \frac{p_x}{p_0} \tag{2.19}$$

$$\varepsilon_v = \frac{-\Delta e}{1+e_0} = \frac{\lambda}{1+e_0} \ln \frac{p_x}{p_0} \tag{2.20}$$

$$\varepsilon_v^e = \frac{\kappa}{1+e_0} \ln \frac{p_x}{p_0} \tag{2.21}$$

$$\varepsilon_v^p = \varepsilon_v - \varepsilon_v^e = \frac{\lambda - \kappa}{1+e_0} \ln \frac{p_x}{p_0} \tag{2.22}$$

where ε_v^e and ε_v^p are the elastic and plastic volumetric strain, respectively.

From Eq. (2.22),

$$\ln p_x = \frac{1+e_0}{\lambda - \kappa} \varepsilon_v^p + \ln p_0 \tag{2.23}$$

Substituting Eq. (2.23) into Eq. (2.18) gives

$$M \ln p + \frac{q}{p} - M \left(\frac{1+e_0}{\lambda - \kappa} \varepsilon_v^p + \ln p_0 \right) = 0 \tag{2.24}$$

Rearranging Eq. (2.24) leads to

$$f = \frac{\lambda - \kappa}{1 + e_0} \ln \frac{p}{p_0} + \frac{\lambda - \kappa}{1 + e_0} \frac{q}{Mp} - \varepsilon_v^p = 0 \tag{2.25}$$

Eq. (2.25) is the yield function of the Cam-clay model, and it can be regarded as $f = f(p, q, \varepsilon_v^p) = 0$. In the plasticity theory, the yield function is usually expressed as $f = f(\sigma_{ij}, H)$, where σ_{ij} = stress tensor and H = hardening parameter. By comparing the two equations, we can conclude that the plastic volumetric strain ε_v^p is the hardening parameter in the Cam-clay model.

Figure 2.9 shows the yield loci of the Cam-clay model in the p–q plane. From Eq. (2.25), we know that the value of ε_v^p is identical on the same yield locus, and can be determined from the results of isotropic compression tests. This implies that when the value of ε_v^p under the isotropic stress state is known, we can obtain the value of ε_v^p under any other possible stress state. In the Cam-clay model, the linear e-ln p relation is adopted. According to author's point of view, this is the second key point in the Cam-clay model.

As shown in Figure 2.9, the value of ε_v^p gradually increases as the yield locus expands. Here, we rewrite Eq. (2.25) as

$$\varepsilon_v^p = \frac{\lambda - \kappa}{1 + e_0} \ln \frac{p}{p_0} + \frac{\lambda - \kappa}{1 + e_0} \frac{q}{Mp} \tag{2.26}$$

where the first term and second term on the right-hand side of Eq. (2.26) indicate the increments of the plastic volumetric strains caused by the increases in p and q/p, respectively (Hata, Ohta, and Yoshitami 1969).

As mentioned earlier, the plastic strain increment vectors are normal to the plastic potential g, i.e.,

$$d\varepsilon_{ij}^p = \Lambda \frac{\partial g}{\partial \sigma_{ij}} \tag{2.27}$$

where Λ is a scalar, and $\partial g / \partial \sigma_{ij}$ shows the direction normal to g.

Because the associated flow rule is adopted in the Cam-clay model, we have

$$d\varepsilon_{ij}^p = \Lambda \frac{\partial f}{\partial \sigma_{ij}} \tag{2.28}$$

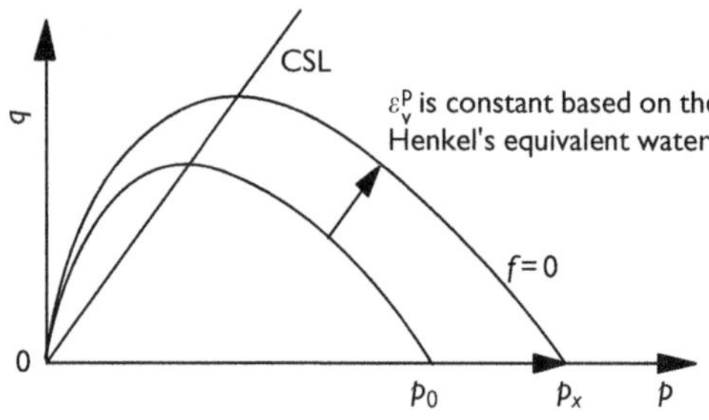

Figure 2.9 The meaning of yield locus in Cam-clay model.

The scalar Λ can be determined from the so-called consistency condition as follows. Since $f = f(p, q, \varepsilon_v^p) = 0$, $df = 0$, i.e.,

$$df = \frac{\partial f}{\partial p}dp + \frac{\partial f}{\partial q}dq + \frac{\partial f}{\partial \varepsilon_v^p}d\varepsilon_v^p = 0 \tag{2.29}$$

where

$$\begin{aligned} d\varepsilon_v^p &= d\varepsilon_1^p + d\varepsilon_2^p + d\varepsilon_3^p = d\varepsilon_{11}^p + d\varepsilon_{22}^p + d\varepsilon_{33}^p \\ &= \Lambda\left(\frac{\partial f}{\partial \sigma_{11}} + \frac{\partial f}{\partial \sigma_{22}} + \frac{\partial f}{\partial \sigma_{33}}\right) = \Lambda\frac{\partial f}{\partial \sigma_{ii}} \quad \text{(summation convention)} \end{aligned} \tag{2.30}$$

Substituting Eq. (2.30) into Eq. (2.29) gives

$$\Lambda = -\frac{(\partial f/\partial p)dp + (\partial f/\partial q)dq}{(\partial f/\partial \varepsilon_v^p)(\partial f/\partial \sigma_{ii})} \tag{2.31}$$

We can calculate $d\varepsilon_{ij}^p$ by substituting Eq. (2.31) into Eq. (2.28).

2.2.4 Determination of $d\varepsilon_{ij}^e$

The elastic strain increment $d\varepsilon_{ij}^e$ can be determined from Hooke's law for isotropic materials. The elastic modulus is related to the swelling index κ from the assumption that soil behaves elastically during the isotropic unloading and reloading.

For isotropic materials, we have

$$\begin{aligned} d\varepsilon_{11}^e &= \frac{1}{E}\{d\sigma_{11} - \nu(d\sigma_{22} + d\sigma_{33})\} \\ &= \frac{1+\nu}{E}d\sigma_{11} - \frac{\nu}{E}(d\sigma_{11} + d\sigma_{22} + d\sigma_{33}) \end{aligned} \tag{2.32}$$

In general, we have

$$d\varepsilon_{ij}^e = \frac{1+\nu}{E}d\sigma_{ij} - \frac{\nu}{E}d\sigma_{mm}\delta_{ij} \tag{2.33}$$

here, $d\sigma_{mm} = d\sigma_{11} + d\sigma_{22} + d\sigma_{33}$ (summation convention), $\delta_{ij} = 1(i = j)$, $0(i \neq j)$ is Keronecker's delta.

From Eq. (2.33),

$$d\varepsilon_v^e = d\varepsilon_{11}^e + d\varepsilon_{22}^e + d\varepsilon_{33}^e = \frac{3(1-2\nu)}{E}dp \tag{2.34}$$

On the other hand, the slope κ of the swelling line leads to (Fig. 2.8),

$$\varepsilon_v^e = \frac{\kappa}{1+e_0}\ln\frac{p}{p_0} \rightarrow d\varepsilon_v^e = \frac{\kappa}{1+e_0}\frac{dp}{p} \tag{2.35}$$

By comparing Eqs (2.34) and (2.35), E is obtained as

$$E = \frac{3(1-2\nu)(1+e_0)}{\kappa}p \tag{2.36}$$

where Poisson's ratio ν is often assumed to be 0 or 0.3 or $\frac{1}{3}$, and it is worth noticing that the elastic modulus E is proportional to the mean stress p.

In conclusion, according to the Cam-clay model, the strain increment $d\varepsilon_{ij}$ can be calculated as follows: $d\varepsilon_{ij} = d\varepsilon_{ij}^{e} + d\varepsilon_{ij}^{p}$; $d\varepsilon_{ij}^{e}$ is calculated using Eq. (2.33), and $d\varepsilon_{ij}^{p}$ is calculated using Eqs (2.28) and (2.31). The model parameters are λ, κ, M, ν, and e_0.

2.2.5 Detailed derivation of $d\varepsilon_{ij}^{p}$

As given in Eq. (2.25), the yield function (= plastic potential) is

$$f = \frac{\lambda-\kappa}{1+e_0}\ln\frac{p}{p_0} + \frac{\lambda-\kappa}{1+e_0}\frac{q}{Mp} - \varepsilon_v^p = 0 \tag{2.25}$$

For example,

$$\frac{\partial f}{\partial \sigma_{11}} = \frac{\partial f}{\partial p}\frac{\partial p}{\partial \sigma_{11}} + \frac{\partial f}{\partial q}\frac{\partial q}{\partial \sigma_{11}} \tag{2.37}$$

In general,

$$\frac{\partial f}{\partial \sigma_{ij}} = \frac{\partial f}{\partial p}\frac{\partial p}{\partial \sigma_{ij}} + \frac{\partial f}{\partial q}\frac{\partial q}{\partial \sigma_{ij}} \tag{2.38}$$

$$\frac{\partial f}{\partial p} = \frac{\lambda-\kappa}{1+e_0}\frac{1}{p} - \frac{\lambda-\kappa}{1+e_0}\frac{1}{M}\frac{q}{p^2} = \frac{\lambda-\kappa}{1+e_0}\frac{1}{Mp}\left(M - \frac{q}{p}\right) \tag{2.39}$$

For example,

$$\frac{\partial p}{\partial \sigma_{11}} = \frac{\partial((\sigma_{11}+\sigma_{22}+\sigma_{33})/3)}{\partial \sigma_{11}} = \frac{1}{3} \tag{2.40}$$

Similarly,

$$\frac{\partial p}{\partial \sigma_{22}} = \frac{1}{3}; \qquad \frac{\partial p}{\partial \sigma_{33}} = \frac{1}{3} \tag{2.41}$$

In general,

$$p = \frac{\sigma_{ii}}{3} \ \text{(summation convention)}, \qquad \frac{\partial p}{\partial \sigma_{ij}} = \frac{\delta_{ij}}{3}, \quad \delta_{ij} = \begin{cases} 1 & (i=j) \\ 0 & (i \neq j) \end{cases} \tag{2.42}$$

$$\frac{\partial f}{\partial q} = \frac{\lambda-\kappa}{1+e_0}\frac{1}{Mp} \tag{2.43}$$

From Eq. (2.5),

$$q = \frac{1}{\sqrt{2}}\sqrt{(\sigma_1-\sigma_2)^2+(\sigma_2-\sigma_3)^2+(\sigma_3-\sigma_1)^2}$$
$$= \frac{1}{\sqrt{2}}\sqrt{(\sigma_{11}-\sigma_{22})^2+(\sigma_{22}-\sigma_{33})^2+(\sigma_{33}-\sigma_{11})^2+3(\sigma_{12}^2+\sigma_{13}^2+\sigma_{21}^2+\sigma_{23}^2+\sigma_{31}^2+\sigma_{32}^2)} \tag{2.44}$$

For example,

$$\frac{\partial q}{\partial \sigma_{11}} = \frac{1}{2\sqrt{2}} \times \frac{2(\sigma_{11}-\sigma_{22})+2(\sigma_{33}-\sigma_{11})\times(-1)}{\sqrt{(\sigma_{11}-\sigma_{22})^2+(\sigma_{22}-\sigma_{33})^2+(\sigma_{33}-\sigma_{11})^2+3(\sigma_{12}^2+\sigma_{13}^2+\sigma_{21}^2+\sigma_{23}^2+\sigma_{31}^2+\sigma_{32}^2)}}$$
$$= \frac{2\sigma_{11}-\sigma_{22}-\sigma_{33}}{2q} = \frac{3(\sigma_{11}-p)}{2q} \tag{2.45}$$

Similarly,

$$\frac{\partial q}{\partial \sigma_{22}} = \frac{3(\sigma_{22}-p)}{2q}, \qquad \frac{\partial q}{\partial \sigma_{33}} = \frac{3(\sigma_{33}-p)}{2q} \tag{2.46}$$

In general,

$$q = \sqrt{\frac{3}{2}(\sigma_{ij}-p\delta_{ij})(\sigma_{ij}-p\delta_{ij})} \tag{2.47}$$

$$\frac{\partial q}{\partial \sigma_{ij}} = \sqrt{\frac{3}{2}}\frac{(\partial(\sigma_{kl}-p\delta_{kl})/\partial\sigma_{ij})(\sigma_{kl}-p\delta_{kl})\times 2}{2\sqrt{(\sigma_{mn}-p\delta_{mn})(\sigma_{mn}-p\delta_{mn})}} = \frac{3\left(\delta_{ik}\delta_{jl}-\frac{1}{3}\delta_{ij}\delta_{kl}\right)(\sigma_{kl}-p\delta_{kl})}{2q}$$
$$= \frac{3(\sigma_{ij}-p\delta_{ij})}{2q} \tag{2.48}$$

Exercise 2.2 Verify Eqs (2.47) and (2.48).

Answer: $(\sigma_{ij}-p\delta_{ij})$ in Eq. (2.47) is the definition of deviatoric stress tensor s_{ij}. The norm of s_{ij} is as follows:

$$\sqrt{s_{ij}s_{ij}} = \sqrt{(\sigma_{ij}-p\delta_{ij})(\sigma_{ij}-p\delta_{ij})}$$
$$= \sqrt{\sigma_{ij}\sigma_{ij}-2p\sigma_{ij}\delta_{ij}+p^2\delta_{ij}\delta_{ij}}$$
$$= \sqrt{\sigma_{ij}\sigma_{ij}-3p^2}$$

Here, $\sigma_{ij}\delta_{ij} = \sigma_{11}\delta_{11} + \sigma_{22}\delta_{22} + \sigma_{33}\delta_{33} = 3p$ and $\delta_{ij}\delta_{ij} = \delta_{11}\delta_{11} + \delta_{22}\delta_{22} + \delta_{33}\delta_{33} = 3$ have been taken into consideration. If the principal stresses are used, we have

$$\begin{aligned}\sqrt{s_{ij}s_{ij}} &= \sqrt{\sigma_{ij}\sigma_{ij} - 3p^2} = \sqrt{\sigma_1^2 + \sigma_2^2 + \sigma_3^2 - 3p^2} \\ &= \sqrt{\sigma_1^2 + \sigma_2^2 + \sigma_3^2 - \left\{\frac{1}{\sqrt{3}}(\sigma_1 + \sigma_2 + \sigma_3)\right\}^2} \\ &= \sqrt{\left|\overrightarrow{\mathrm{OS}}\right|^2 - \left|\overrightarrow{\mathrm{OP}}\right|^2} = \left|\overrightarrow{\mathrm{PS}}\right| \quad \text{(see Eq. (2.3) and Fig. 2.2(a))}\end{aligned}$$

From Eq. (2.3), q can be expressed as

$$q = \sqrt{\frac{3}{2}}\left|\overrightarrow{\mathrm{PS}}\right| = \sqrt{\frac{3}{2}}\sqrt{s_{ij}s_{ij}} = \sqrt{\frac{3}{2}}\sqrt{(\sigma_{ij} - p\delta_{ij})(\sigma_{ij} - p\delta_{ij})} \tag{2.47}$$

Next,

$$\frac{\partial q}{\partial \sigma_{ij}} = \sqrt{\frac{3}{2}}\frac{\partial\{(\sigma_{kl} - p\delta_{kl})(\sigma_{kl} - p\delta_{kl})\}/\partial\sigma_{ij}}{2\sqrt{(\sigma_{mn} - p\delta_{mn})(\sigma_{mn} - p\delta_{mn})}}$$

where

$$\begin{aligned}&\frac{\partial}{\partial\sigma_{ij}}\{(\sigma_{kl} - p\delta_{kl})(\sigma_{kl} - p\delta_{kl})\} \\ &= \left(\frac{\partial\sigma_{kl}}{\partial\sigma_{ij}} - \frac{\partial p}{\partial\sigma_{ij}}\delta_{kl}\right)(\sigma_{kl} - p\delta_{kl}) + (\sigma_{kl} - p\delta_{kl})\left(\frac{\partial\sigma_{kl}}{\partial\sigma_{ij}} - \frac{\partial p}{\partial\sigma_{ij}}\delta_{kl}\right) \\ &= 2 \times \left(\delta_{ik}\delta_{jl} - \frac{1}{3}\delta_{ij}\delta_{kl}\right)(\sigma_{kl} - p\delta_{kl}) \\ &= 2 \times \left(\delta_{ik}\delta_{jl}\sigma_{kl} - \frac{1}{3}\delta_{ij}\delta_{kl}\sigma_{kl} - p\delta_{ik}\delta_{jl}\delta_{kl} + \frac{p}{3}\delta_{ij}\delta_{kl}\delta_{kl}\right) \\ &= 2 \times (\sigma_{ij} - p\delta_{ij} - p\delta_{ij} + p\delta_{ij}) = 2 \times (\sigma_{ij} - p\delta_{ij})\end{aligned}$$

here, $\sigma_{kl}\delta_{kl} = 3p$ and $\delta_{kl}\delta_{kl} = 3$ are also taken into consideration.

$$\begin{aligned}\therefore \frac{\partial q}{\partial\sigma_{ij}} &= \sqrt{\frac{3}{2}}\frac{2(\sigma_{ij} - p\delta_{ij})}{2\sqrt{(\sigma_{mn} - p\delta_{mn})(\sigma_{mn} - p\delta_{mn})}} \\ &= \frac{3(\sigma_{ij} - p\delta_{ij})}{2q}\end{aligned} \tag{2.48}$$

From Eqs (2.38), (2.39), (2.42), (2.43), and (2.48),

$$\begin{aligned}\frac{\partial f}{\partial\sigma_{ij}} &= \frac{\lambda - \kappa}{1 + e_0}\frac{1}{M}\frac{1}{p}\left(M - \frac{q}{p}\right)\frac{\delta_{ij}}{3} + \frac{\lambda - \kappa}{1 + e_0}\frac{1}{M}\frac{1}{p}\frac{3\left(\sigma_{ij} - p\delta_{ij}\right)}{2q} \\ &= \frac{\lambda - \kappa}{1 + e_0}\frac{1}{Mp}\left\{\frac{1}{3}\left(M - \frac{q}{p}\right)\delta_{ij} + \frac{3\left(\sigma_{ij} - p\delta_{ij}\right)}{2q}\right\}\end{aligned} \tag{2.49}$$

Eq. (2.31) is rewritten here:

$$\Lambda = -\frac{(\partial f/\partial p)dp + (\partial f/\partial q)dq}{(\partial f/\partial \varepsilon_{\mathrm{v}}^{\mathrm{p}})(\partial f/\partial \sigma_{ii})} \tag{2.31}$$

where from Eq. (2.25)

$$\frac{\partial f}{\partial \varepsilon_{\mathrm{v}}^{\mathrm{p}}} = -1 \tag{2.50}$$

and from Eq. (2.49)

$$\begin{aligned}\frac{\partial f}{\partial \sigma_{ii}} &= \frac{\lambda-\kappa}{1+e_0}\frac{1}{Mp}\left\{\left(M-\frac{q}{p}\right)+\frac{3(\sigma_{11}-p)}{2q}+\frac{3(\sigma_{22}-p)}{2q}+\frac{3(\sigma_{33}-p)}{2q}\right\}\\ &= \frac{\lambda-\kappa}{1+e_0}\frac{1}{Mp}\left(M-\frac{q}{p}\right)\end{aligned} \tag{2.51}$$

Therefore, Λ can be obtained from Eqs (2.39), (2.43), (2.50), and (2.51),

$$\begin{aligned}\Lambda &= \frac{(\lambda-\kappa)/(1+e_0)(1/Mp)(M-(q/p))dp + ((\lambda-\kappa)/(1+e_0))(1/Mp)dq}{(\lambda-\kappa)/(1+e_0)(1/Mp)(M-(q/p))}\\ &= dp + \frac{dq}{(M-(q/p))}\end{aligned} \tag{2.52}$$

$$\begin{aligned}\therefore d\varepsilon_{ij}^{\mathrm{p}} &= \Lambda\frac{\partial f}{\partial \sigma_{ij}}\\ &= \frac{\lambda-\kappa}{1+e_0}\frac{1}{Mp}\left\{\frac{1}{3}\left(M-\frac{q}{p}\right)\delta_{ij}+\frac{3(\sigma_{ij}-p\delta_{ij})}{2q}\right\}\left\{dp+\frac{dq}{(M-(q/p))}\right\}\end{aligned} \tag{2.53}$$

Therefore, Eq. (2.53) is a general equation for calculating the plastic strain increment $d\varepsilon_{ij}^{\mathrm{p}}$. As an example, let us discuss the plastic volumetric strain increment $d\varepsilon_{\mathrm{v}}^{\mathrm{p}}$ and the plastic deviatoric strain increment $d\varepsilon_{\mathrm{d}}^{\mathrm{p}}$. From Eq. (2.53), we have

$$d\varepsilon_{11}^{\mathrm{p}} = \frac{\lambda-\kappa}{1+e_0}\frac{1}{Mp}\left\{\frac{1}{3}\left(M-\frac{q}{p}\right)+\frac{3(\sigma_{11}-p)}{2q}\right\}\left\{dp+\frac{dq}{(M-(q/p))}\right\} \tag{2.54}$$

$$d\varepsilon_{22}^{\mathrm{p}} = \frac{\lambda-\kappa}{1+e_0}\frac{1}{Mp}\left\{\frac{1}{3}\left(M-\frac{q}{p}\right)+\frac{3(\sigma_{22}-p)}{2q}\right\}\left\{dp+\frac{dq}{(M-(q/p))}\right\} \tag{2.55}$$

$$d\varepsilon_{33}^{\mathrm{p}} = \frac{\lambda-\kappa}{1+e_0}\frac{1}{Mp}\left\{\frac{1}{3}\left(M-\frac{q}{p}\right)+\frac{3(\sigma_{33}-p)}{2q}\right\}\left\{dp+\frac{dq}{(M-(q/p))}\right\} \tag{2.56}$$

Because $d\varepsilon_{\mathrm{v}}^{\mathrm{p}} = d\varepsilon_{11}^{\mathrm{p}} + d\varepsilon_{22}^{\mathrm{p}} + d\varepsilon_{33}^{\mathrm{p}}$ and $\sigma_{11}+\sigma_{22}+\sigma_{33} = 3p$,

$$d\varepsilon_{\mathrm{v}}^{\mathrm{p}} = \frac{\lambda-\kappa}{1+e_0}\frac{1}{Mp}\left(M-\frac{q}{p}\right)\left\{dp+\frac{dq}{(M-(q/p))}\right\} \tag{2.57}$$

$d\varepsilon_v^p$ can also be calculated from Eq. (2.26), i.e.,

$$\begin{aligned} d\varepsilon_v^p &= \frac{\partial \varepsilon_v^p}{\partial p} dp + \frac{\partial \varepsilon_v^p}{\partial q} dq \\ &= \left(\frac{\lambda - \kappa}{1 + e_0} \frac{1}{p} - \frac{\lambda - \kappa}{1 + e_0} \frac{1}{M} \frac{q}{p^2} \right) dp + \frac{\lambda - \kappa}{1 + e_0} \frac{1}{M} \frac{1}{p} dq \\ &= \frac{\lambda - \kappa}{1 + e_0} \frac{1}{Mp} \left(M - \frac{q}{p} \right) dp + \frac{\lambda - \kappa}{1 + e_0} \frac{1}{Mp} dq \\ &= \frac{\lambda - \kappa}{1 + e_0} \frac{1}{Mp} \left(M - \frac{q}{p} \right) \left\{ dp + \frac{dq}{(M - (q/p)} \right\} \end{aligned} \quad (2.58)$$

As given in Eq. (2.14), $(d\varepsilon_v^p / d\varepsilon_d^p) = M - (q/p)$, $d\varepsilon_d^p$ can be directly calculated as follows:

$$d\varepsilon_d^p = \frac{\lambda - \kappa}{1 + e_0} \frac{1}{Mp} \left\{ dp + \frac{dq}{(M - (q/p))} \right\} \quad (2.59)$$

Therefore, we have the following simple matrix form from Eqs (2.57) and (2.59).

$$\begin{Bmatrix} d\varepsilon_v^p \\ d\varepsilon_d^p \end{Bmatrix} = \frac{\lambda - \kappa}{1 + e_0} \frac{1}{Mp} \begin{bmatrix} (M - (q/p)) & 1 \\ 1 & \dfrac{1}{(M - (q/p))} \end{bmatrix} \begin{Bmatrix} dp \\ dq \end{Bmatrix} \quad (2.60)$$

Based on the discussion so far, it becomes clear that the Cam-clay model has two key elements. First, the plastic volumetric strain is taken as the hardening parameter, and its value is determined from the linear e-log p relation under isotropic stress states (Eq. (2.25) or Eq. (2.26)). Second, the relation between the stress ratio and the strain increment ratio (Eqs (2.14)). Equations (2.26), and (2.14) are used during the derivation of $d\varepsilon_v^p$ and $d\varepsilon_d^p$.

Exercise 2.3 Using the Cam-clay model, calculate the stress–strain relation ($q/p - \varepsilon_d$ relation, $\varepsilon_v - \varepsilon_d$ relation, $\sigma_1'/\sigma_3' - \varepsilon_1$, ε_3 relation, and $\varepsilon_v - \varepsilon_1$ relation) and the effective stress path (q–p relation and $(\sigma_1' - \sigma_3')/2 \sim (\sigma_1' + \sigma_3')/2$ relation) under the following test conditions. Here, σ_1, σ_3, and σ_m are the major, minor, and mean principal stress in terms of total stress, and σ_1', σ_3', and p are the major, minor, and mean principal stress in terms of effective stress while in this book σ_{ij} means the effective stress.

(a) Drained triaxial compression tests and drained triaxial extension tests under $p = 196$ kPa, $\sigma_3' = 196$ kPa, and $\sigma_1' = 196$ kPa
(b) Undrained triaxial compression tests and undrained triaxial extension tests under $\sigma_m = 196$ kPa, $\sigma_3 = 196$ kPa, and $\sigma_1 = 196$ kPa

The values of the model parameters are as follows: $C_c/(1+e_0) = 11.7\%$, $C_s/(1+e_0) = 1.6\%$, $\phi = 34°$, and $\nu = 0.3$ (e_0 is the initial void ratio and ν is Poisson's ratio).

Answer: As mentioned, the model parameters are λ, κ, M, e_0, and ν. λ and κ are the slopes of the compression line and swelling line in the e-ln p plane, and are related to the compression index C_c and the swelling index C_s.

$$\lambda = \frac{-\Delta e}{\Delta(\ln p)} = \frac{-\Delta e}{\Delta(\log p / \log e)} = \frac{-\Delta e}{\Delta \log p} \log e = 0.434 C_c \quad (2.61)$$

Similarly,

$$\kappa = 0.434C_s \tag{2.62}$$

Under triaxial compression condition, the stress ratio $q/p(= M)$ at the critical state can be expressed by

$$M = \left[\frac{q}{p}\right]_{cs} = \left[\frac{\sigma_1 - \sigma_3}{(\sigma_1 + 2\sigma_3)/3}\right]_{cs} = \left[\frac{3(\sigma_1/\sigma_3 - 1)}{\sigma_1/\sigma_3 + 2}\right]_{cs} \tag{2.63}$$

and

$$\left(\frac{\sigma_1}{\sigma_3}\right)_{cs} = \frac{1 + \sin\phi}{1 - \sin\phi} \tag{2.64}$$

Substituting the above equation into Eq. (2.63) gives

$$M = \frac{6\sin\phi}{3 - \sin\phi} \tag{2.65}$$

(a) Prediction of stress–strain relation under drained condition Since the effective stresses are known under drained condition, the strains can be calculated by imposing the stress increments from the initial stress condition. As an example, Table 2.1 shows the calculated results of the triaxial compression test under $\sigma_3' = 196$ kPa.

Let us comment on Table 2.1.

In column ①, when the number of steps is small, the increments of σ_1' are initially large and then reduced as the stress state approaches failure. This arrangement of step sizes generally leads to more accurate calculation. Eq. (2.33) is used to obtain the columns ⑦ and ⑧. The elastic modulus E is calculated using Eq. (2.36). Columns ⑨ and ⑩ are calculated using Eqs (2.54) and (2.56).

Figure 2.10 shows the predicted results of the triaxial compression test under $\sigma_3' = 196$ kPa from Table 2.1. The solid curves are the model prediction and the black marks are the calculated values in Table 2.1. The same black marks are also used in Figures 2.11–2.13 and 2.18–2.20.

Figure 2.11 shows the stress path of the triaxial compression test under $\sigma_3' = 196$ kPa. Since it is a drained test, the effective stress path is the same as the total stress path.

(b) Prediction of stress–strain relation and stress path under undrained condition The undrained condition can be expressed as $d\varepsilon_v = 0$, or,

$$d\varepsilon_v = d\varepsilon_v^e + d\varepsilon_v^p = 0 \tag{2.66}$$

Substituting Eqs (2.35) and (2.57) into Eq. (2.66) gives

$$\frac{\kappa}{1 + e_0} \cdot \frac{dp}{p} + \frac{\lambda - \kappa}{1 + e_0} \cdot \frac{1}{Mp}\{(M - (q/p))dp + dq\} = 0 \tag{2.67}$$

An undrained triaxial extension test under $\sigma_m = 196$ kPa is taken as an example to explain how to predict the stress–strain relation and stress path under undrained condition.

Table 2.1 Predicted results using original Cam-clay model for drained triaxial compression test at $\sigma'_3 = 196\,\text{kPa}$

① σ'_1 (kPa)	② σ'_3 (kPa)	③ p (kPa)	④ q (kPa)	⑤ dp (kPa)	⑥ dq (kPa)	⑦ $d\varepsilon_1^e$ (%)	⑧ $d\varepsilon_3^e$ (%)	⑨ $d\varepsilon_1^p$ (%)	⑩ $d\varepsilon_3^p$ (%)	⑪ ε_1 (%)	⑫ ε_3 (%)	⑬ ε_d (%)	⑭ ε_v (%)
196	196	196.00	0							0	0	0	0
245	196	212.33	49	16.33	49.0	0.134	−0.040	1.228	−0.106	1.362	−0.146	1.005	1.071
294	196	228.67	98	16.33	49.0	0.124	−0.037	1.251	−0.176	2.737	−0.359	2.064	2.019
333	196	241.67	137	13.00	39.0	0.094	−0.028	1.033	−0.188	3.863	−0.575	2.959	2.713
372	196	254.67	176	13.00	39.0	0.089	−0.027	1.084	−0.241	5.037	−0.843	3.920	3.351
402	196	264.67	206	10.00	30.0	0.064	−0.019	0.854	−0.215	5.955	−1.077	4.688	3.802
431	196	274.33	235	9.67	29.0	0.062	−0.019	0.909	−0.254	6.926	−1.349	5.517	4.227
461	196	284.33	265	10.00	30.0	0.060	−0.018	0.981	−0.301	7.967	−1.669	6.424	4.630
490	196	294.00	294	9.67	29.0	0.058	−0.017	1.079	−0.360	9.104	−2.046	7.433	5.012
519	196	303.67	323	9.67	29.0	0.056	−0.017	1.214	−0.437	10.374	−2.500	8.582	5.375
549	196	313.67	353	10.00	30.0	0.054	−0.016	1.409	−0.542	11.837	−3.058	9.930	5.721
578	196	323.33	382	9.67	29.0	0.053	−0.016	1.709	−0.700	13.598	−3.774	11.581	6.050
608	196	333.33	412	10.00	30.0	0.051	−0.015	2.221	−0.963	15.871	−4.753	13.749	6.365
627	196	339.67	431	6.33	19.0	0.033	−0.010	1.882	−0.846	17.786	−5.608	15.596	6.569
647	196	346.33	451	6.67	20.0	0.033	−0.010	2.626	−1.221	20.445	−6.839	18.189	6.767
666	196	352.67	470	6.33	19.0	0.032	−0.010	4.477	−2.149	24.954	−8.997	22.634	6.959
686	196	359.33	490	6.67	20.0	0.032	−0.009	16.907	−8.366	41.893	−17.37	39.511	7.147

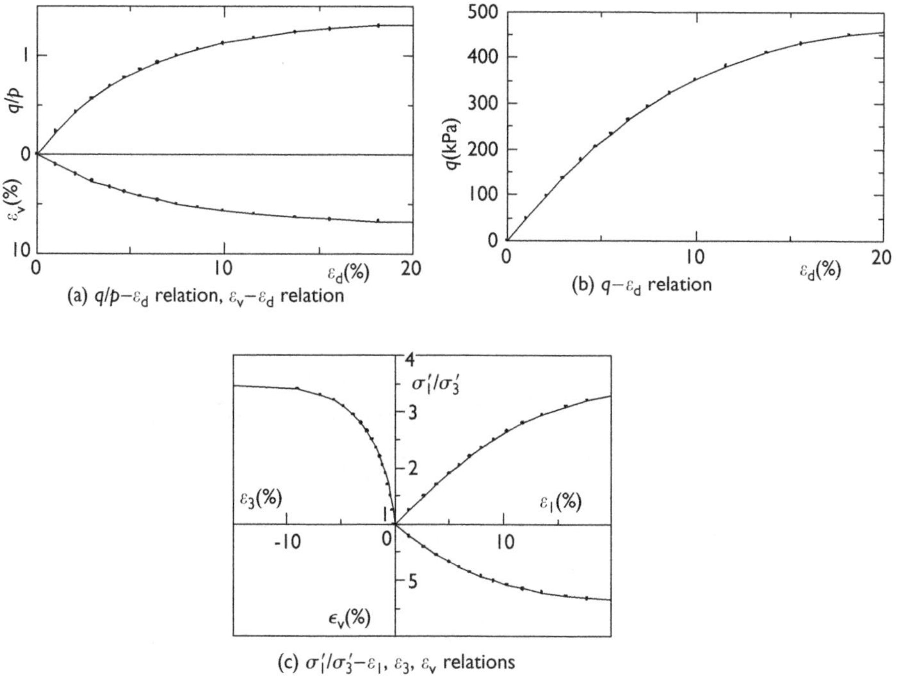

Figure 2.10 Prediction of stress versus strain relation in drained triaxial compression test by original Cam-clay model ($\sigma'_3 = 196\,\text{kPa}$).

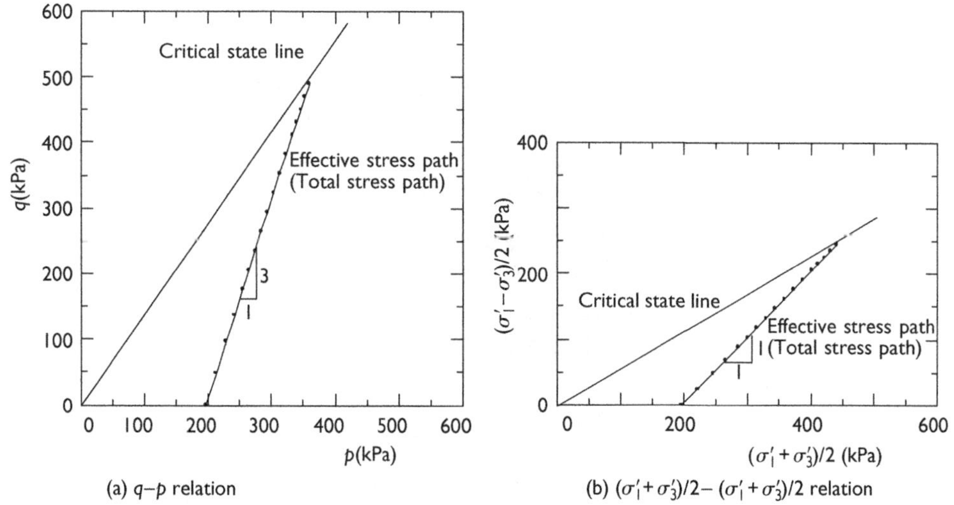

Figure 2.11 Prediction of stress paths in drained triaxial compression test by original Cam-clay model ($\sigma'_3 = 196\,\text{kPa}$).

Table 2.2 Predicted results using original Cam-clay model for undrained triaxial extension test under σ_m = 196 kPa

① σ_3 (kPa)	② σ_1 (kPa)	③ σ_m (kPa)	④ du (kPa)	⑤ u (kPa)	⑥ σ'_3 (kPa)	⑦ σ'_1 (kPa)	⑧ p (kPa)	⑨ q (kPa)	⑩ dp (kPa)	⑪ dq (kPa)	⑫ $d\varepsilon_3^e$ (%)	⑬ $d\varepsilon_1^e$ (%)	⑭ $d\varepsilon_3^p$ (%)	⑮ $d\varepsilon_1^p$ (%)	⑯ ε_3 (%)	⑰ ε_1 (%)	⑱ ε_d (%)	⑲ ε_v (%)
196.00	196.00	196		0.00	196.00	196.00	196.00	0.00							0.000	0.000	0.000	0.000
186.20	200.90	196	9.692	9.69	176.51	191.21	186.31	14.70	−9.69	14.70	−0.052	0.008	−0.016	0.026	−0.068	0.034	0.068	0.001
176.40	205.80	196	10.252	19.94	156.46	185.86	176.06	29.40	−10.25	14.70	−0.055	0.007	−0.021	0.032	−0.144	0.073	0.145	0.002
166.60	210.70	196	10.961	30.91	135.70	179.80	165.10	44.10	−10.96	14.70	−0.060	0.007	−0.028	0.039	−0.232	0.119	0.234	0.006
156.80	215.60	196	11.898	42.80	114.00	172.80	153.20	58.80	−11.90	14.70	−0.066	0.006	−0.041	0.051	−0.340	0.176	0.344	0.013
152.88	217.56	196	5.028	47.83	105.05	169.73	148.17	64.68	−5.03	5.88	−0.028	0.002	−0.018	0.022	−0.386	0.200	0.391	0.014
147.00	220.50	196	8.050	55.88	91.12	164.62	140.12	73.50	−8.05	8.82	−0.045	0.002	−0.038	0.042	−0.469	0.244	0.475	0.020
141.12	223.44	196	8.783	64.67	76.46	158.78	131.34	82.32	−8.78	8.82	−0.049	0.001	−0.055	0.055	−0.573	0.300	0.582	0.028
137.20	225.40	196	6.391	71.06	66.14	154.34	124.94	88.20	−6.39	5.88	−0.035	0.000	−0.048	0.044	−0.656	0.345	0.667	0.033
133.28	227.36	196	7.011	78.07	55.21	149.29	117.93	94.08	−7.01	5.88	−0.039	−0.001	−0.070	0.060	−0.765	0.403	0.779	0.042
127.40	230.30	196	12.265	90.33	37.07	139.97	105.67	102.90	−12.26	8.82	−0.069	−0.006	−0.256	0.187	−1.089	0.584	1.115	0.079
125.44	231.28	196	4.983	95.31	30.13	135.97	100.69	105.84	−4.98	2.94	−0.026	−0.004	−0.117	0.079	−1.232	0.660	1.261	0.087
123.48	232.26	196	5.750	101.06	22.42	131.20	94.94	108.78	−5.75	2.94	−0.030	−0.006	−0.222	0.138	−1.483	0.791	1.516	0.100
121.52	233.24	196	7.086	108.15	13.37	125.09	87.85	111.72	−7.09	2.94	−0.035	−0.010	−0.760	0.420	−2.278	1.201	2.319	0.124
121.13	233.44	196	1.877	110.03	11.10	123.41	85.97	112.31	−1.88	0.59	−0.008	−0.003	−0.247	0.132	−2.534	1.330	2.576	0.126
120.74	233.63	196	2.111	112.14	8.60	121.49	83.86	112.90	−2.11	0.59	−0.009	−0.004	−0.718	0.369	−3.261	1.695	3.304	0.129
120.54	233.73	196	1.215	113.35	7.19	120.38	82.65	113.19	−1.22	0.29	−0.005	−0.003	−2.512	1.262	−5.778	2.954	5.821	0.130
120.52	233.74	196	0.133	113.49	7.03	120.25	82.51	113.22	−0.13	0.03	−0.001	0.000	−0.602	0.301	−6.381	3.255	6.424	0.130
120.51	233.74	196	0.040	113.53	6.99	120.22	82.47	113.23	−0.04	0.01	0.000	0.000	−0.310	0.155	−6.691	3.410	6.734	0.130
120.51	233.75	196	0.054	113.58	6.93	120.17	82.42	113.24	−0.05	0.01	0.000	0.000	−8.001	4.001	−14.69	7.411	14.735	0.130

Since σ_m is constant, we have

$$dp = d(\sigma_m - u) = -du \tag{2.68}$$

$$dq = d(\sigma_1' - \sigma_3') = d(\sigma_1 - \sigma_3) = d\sigma_1 - d\sigma_3 \tag{2.69}$$

where u is the excess pore water pressure.

Substituting Eq. (2.68) into Eq. (2.67) gives

$$du = \frac{dq}{M + (M\kappa/(\lambda - \kappa)) - (q/(\sigma_m - u))} \tag{2.70}$$

The difference in the prediction between drained and undrained conditions is whether the effective stress is given or not. The excess pore water pressure u can be calculated using Eq. (2.70) from the given total stress, and the effective stress can then be calculated from the values of the total stress and the excess pore water pressure. Once the effective stress is known, the calculation follows the same procedure as in the drained test.

Table 2.2 shows the predicted results of the undrained triaxial extension test under $\sigma_m = 196$ kPa using the original Cam-clay model.

Figures 2.12 and 2.13 show the predicted results in Table 2.2. In Figure 2.12, we note that there is a very small volumetric strain during shearing. This small volumetric strain is due to the numerical error. The more the steps we use, the smaller the value of $d\varepsilon_v$ should be. Strictly speaking, we should have $d\varepsilon_v = 0$ when the step size approaches zero.

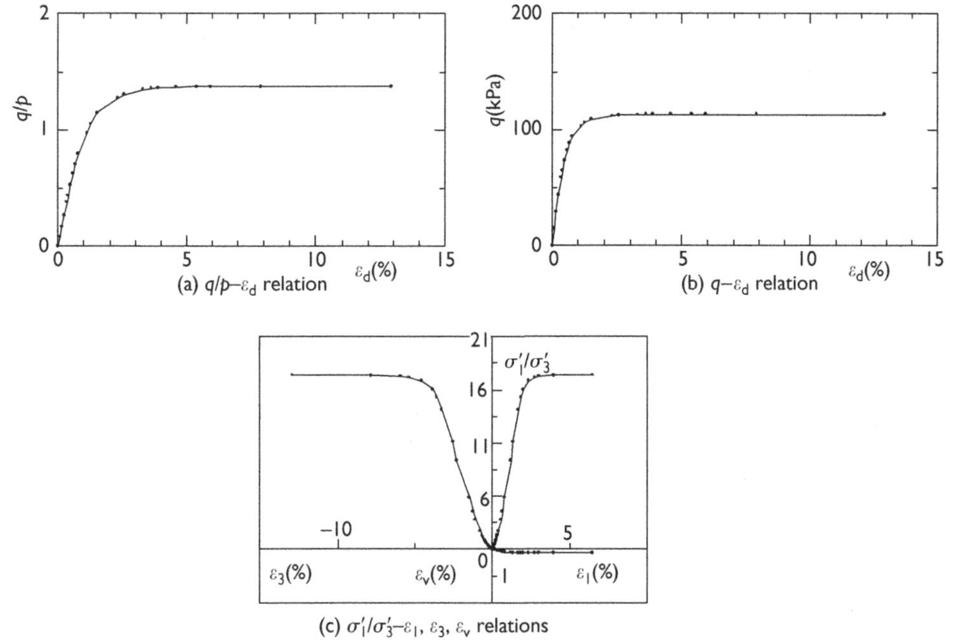

Figure 2.12 Prediction of stress versus strain relation in undrained triaxial extension test by original Cam-clay model ($\sigma_m = 196$ kPa).

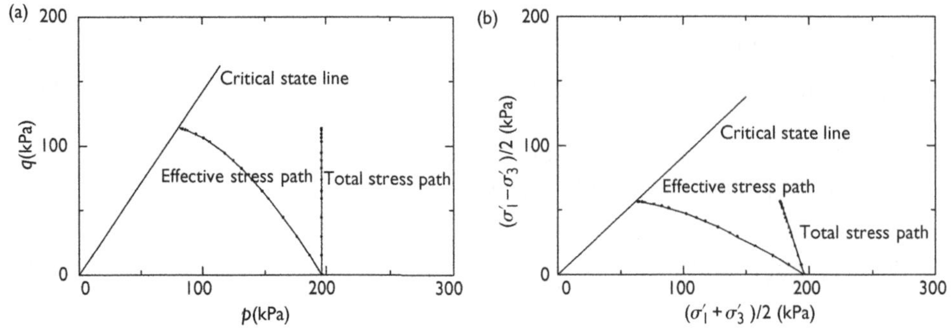

Figure 2.13 Prediction of stress paths in undrained triaxial extension test by original Cam-clay model ($\sigma_m = 196$ kPa).

2.3 Modified Cam-clay model

The original Cam-clay model was proposed in 1963. As shown in Figure 2.5, the plastic potential curve is not normal to the p-axis, which results in plastic shear strain during the isotropic compression loading. This contradicts the experimental results. As is well known, for isotropic material only isotropic deformation occurs when subjected to isotropic compression. In order to avoid this shortcoming, the modified Cam-clay model was proposed in 1968 by Roscoe and Burland. In the modified Cam-clay model, the plastic potential curve as well as the yield locus is normal to the p-axis, and is assumed to be elliptic for simplicity, although a different energy dissipation equation than that of the original Cam-clay model is used in the original paper of Roscoe and Burland (1968). Therefore, the only difference between the original and the modified Cam-clay model is the energy equations used in the two models, which leads to the difference in the relation between the stress ratio and the strain increment ratio. In the following sections, a detailed description of the modified Cam-clay model is given.

2.3.1 *Principal direction of plastic strain increment and adopted stress and strain variables*

These are all the same as those used in the original Cam-clay model (see Section 2.2.1).

2.3.2 *Determination of plastic potential and yield functions*

The equation of plastic work is rewritten as

$$dW^{\mathrm{p}} = \sigma_1 d\varepsilon_1^{\mathrm{p}} + \sigma_2 d\varepsilon_2^{\mathrm{p}} + \sigma_3 d\varepsilon_3^{\mathrm{p}} = p d\varepsilon_{\mathrm{v}}^{\mathrm{p}} + q d\varepsilon_{\mathrm{d}}^{\mathrm{p}} \tag{2.12}$$

In the modified Cam-clay model, the energy dissipation equation is assumed to be

$$dW^{\mathrm{p}} = p\sqrt{(d\varepsilon_{\mathrm{v}}^{\mathrm{p}})^2 + (M d\varepsilon_{\mathrm{d}}^{\mathrm{p}})^2} \tag{2.71}$$

That is,

$$dW^{\mathrm{p}} = p\,d\varepsilon_{\mathrm{v}}^{\mathrm{p}} + q\,d\varepsilon_{\mathrm{d}}^{\mathrm{p}} = p\sqrt{(d\varepsilon_{\mathrm{v}}^{\mathrm{p}})^2 + (Md\varepsilon_{\mathrm{d}}^{\mathrm{p}})^2} \tag{2.72}$$

Rearranging Eq. (2.72) gives

$$\frac{d\varepsilon_{\mathrm{v}}^{\mathrm{p}}}{d\varepsilon_{\mathrm{d}}^{\mathrm{p}}} = \frac{M^2 - (q/p)^2}{2(q/p)} = \frac{M^2p^2 - q^2}{2pq} \tag{2.73}$$

or

$$\frac{q}{p} = \sqrt{M^2 + \left(\frac{d\varepsilon_{\mathrm{v}}^{\mathrm{p}}}{d\varepsilon_{\mathrm{d}}^{\mathrm{p}}}\right)^2} - \frac{d\varepsilon_{\mathrm{v}}^{\mathrm{p}}}{d\varepsilon_{\mathrm{d}}^{\mathrm{p}}}$$

This equation represents the relation between the stress ratio and the plastic strain increment ratio used in the modified Cam-clay model, and is depicted as Figure 2.14.

Next, combining the relation between the direction of the plastic strain increment vector and the stress ratio (Eq. (2.73)) with the normality condition (Eq. (2.11)), we can obtain the plastic potential function in the p–q plane. Substituting Eq. (2.11) into Eq. (2.73) gives

$$\frac{dq}{dp} + \frac{M^2 - (q/p)^2}{2(q/p)} = 0 \tag{2.74}$$

Exercise 2.4 Solve the ordinary differential equation (2.74).

Answer:

Letting $\frac{q}{p} = x$ gives

$$q = px$$

$$\frac{dq}{dp} = \frac{d(px)}{dp} = x + p\frac{dx}{dp}$$

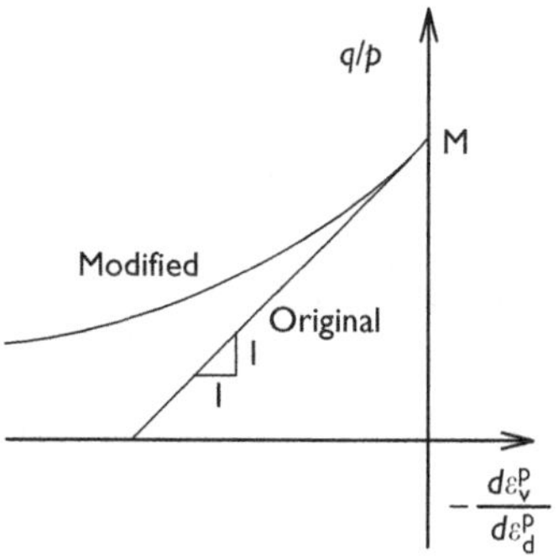

Figure 2.14 Stress ratio versus plastic strain increment ratio adopted in modified Cam-clay model.

So, Eq. (2.74) can be written as

$$\int \frac{2x\,dx}{x^2+M^2} + \int \frac{dp}{p} = C' \text{ (C': an integral constant)}$$

$$\ln(x^2+M^2) + \ln p = \ln C \quad (C' = \ln C)$$

$$(x^2+M^2)p = C$$

$$\left\{\left(\frac{q}{p}\right)^2 + M^2\right\} p = C$$

$\therefore q^2 + M^2p^2 - Cp = 0$ (C: an integral constant)
Therefore, the plastic potential function in the modified Cam-clay model can be written as

$$g = q^2 + M^2p^2 - Cp = 0 \tag{2.75}$$

where C is an integral constant. Figure 2.15 shows the relation between the plastic strain increment vectors indicated by Eq. (2.73) and the plastic potential defined by Eq. (2.75). From Eq. (2.73), we know that $d\varepsilon_d^p = 0$ when $q/p = 0$, and $d\varepsilon_v^p = 0$ when $q/p = M$. The latter also defines the so-called Critical State Line (CSL). The former indicates that the shortcoming of the original Cam-clay model along the isotropic stress path has been overcome. From Eq. (2.75) and Figure 2.15, we also know that the plastic potential is elliptic.

In the modified Cam-clay model, the associated flow rule is adopted, i.e., the yield function f is equal to the plastic potential function g. Hence, the yield function f takes the form

$$f = q^2 + M^2p^2 - Cp = 0 \tag{2.76}$$

Let $p = p_x$ when $q = 0$. Substituting p_x into Eq. (2.76) results in $C = M^2p_x$. Therefore, we eventually have

$$f = q^2 + M^2p^2 - M^2p_xp = 0 \tag{2.77}$$

Figure 2.16 shows the yield locus of the modified Cam-clay model.

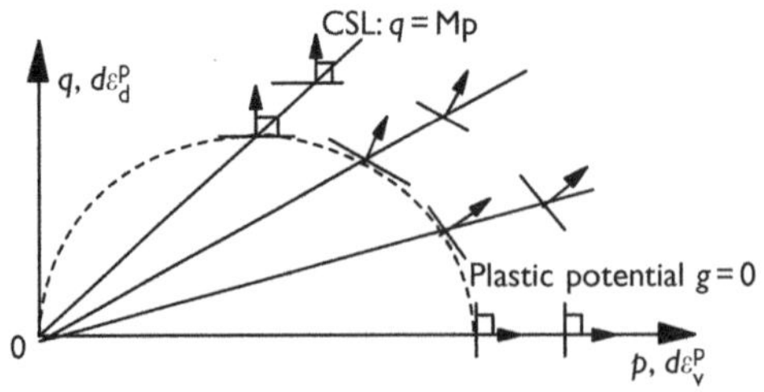

Figure 2.15 Plastic potential and plastic strain increment vectors in modified Cam-clay model.

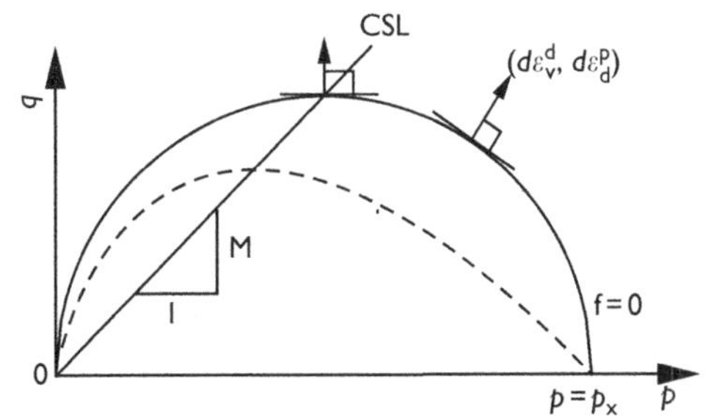

Figure 2.16 Yield locus of modified Cam-clay model (dotted curve is the yield locus of original Cam-clay model).

2.3.3 Determination of strain-hardening rule

The strain-hardening rule used in the modified Cam-clay model is the same as that used in the original Cam-clay model. Rearranging Eq. (2.77) leads to

$$q^2 + M^2p^2 = M^2p^2 \times \frac{p_x}{p} \tag{2.78}$$

Taking the natural logarithms of the two sides of Eq. (2.78) gives

$$\ln(q^2 + M^2p^2) = \ln M^2p^2 + \ln p_x - \ln p \tag{2.79}$$

$$\ln\left(1 + \frac{q^2}{M^2p^2}\right) = \ln p_x - \ln p \tag{2.80}$$

Substituting Eq. (2.23) into Eq. (2.80) gives

$$\ln\left(1 + \frac{q^2}{M^2p^2}\right) = \frac{1+e_0}{\lambda-\kappa}\varepsilon_v^p + \ln p_0 - \ln p \tag{2.81}$$

Rearranging the equation above leads to the following yield function:

$$f = \frac{\lambda-\kappa}{1+e_0}\ln\frac{p}{p_0} + \frac{\lambda-\kappa}{1+e_0}\ln\left(1 + \frac{q^2}{M^2p^2}\right) - \varepsilon_v^p = 0 \tag{2.82}$$

This equation defines the yield function of the modified Cam-clay model. As the original Cam-clay model, the modified Cam-clay model uses stress invariants p and q as the stress variables and the plastic volumetric strain ε_v^p as the hardening parameter.

As shown in Figure 2.17, the yield loci of the modified Cam-clay model are ellipses in the p–q plane. From Eq. (2.82), we know that the value of ε_v^p is identical along the same yield locus, and the value of ε_v^p gradually increases as the yield loci expand. Here, we rewrite Eq. (2.82) as

$$\varepsilon_v^p = \frac{\lambda-\kappa}{1+e_0}\ln\frac{p}{p_0} + \frac{\lambda-\kappa}{1+e_0}\ln\left(1 + \frac{q^2}{M^2p^2}\right) \tag{2.83}$$

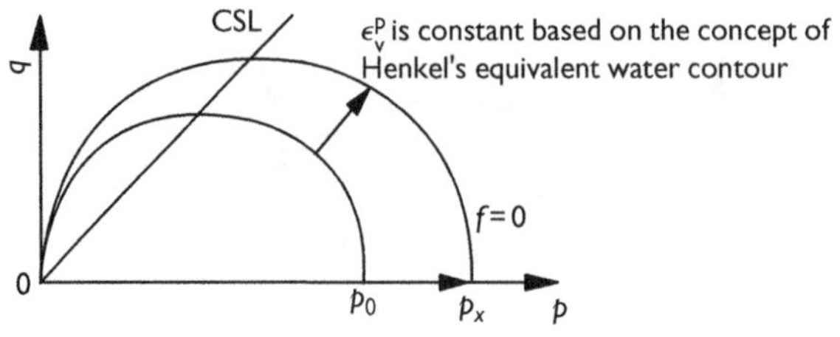

Figure 2.17 The meaning of yield locus in modified Cam-clay model.

where the first term and the second term on the right-hand side of Eq. (2.83) indicate the increments of the plastic volumetric strain induced by the increases in p and q/p, respectively. At the critical state, the second term on the right-hand side of Eq. (2.26) for the original Cam-clay model is $(\lambda-\kappa)/(1+e_0)$, while that of Eq. (2.83) for the modified Cam-clay model is $(\lambda-\kappa)/(1+e_0)\times\ln 2=(\lambda-\kappa)/(1+e_0)\times 0.693$. The latter is about 70% of the former, i.e., for constant p the predicted plastic volumetric strain by modified Cam-clay model at the critical state is about 70% of that by original Cam-clay model.

2.3.4 Determination of $d\varepsilon_{ij}^{e}$

The method for calculating $d\varepsilon_{ij}^{e}$ is the same as that in the original Cam-clay model. It can be calculated using Eqs (2.33) and (2.36).

2.3.5 Detailed derivation of $d\varepsilon_{ij}^{p}$

As mentioned earlier, the difference between the two models is only the yield function (or the plastic potential). Rewriting Eq. (2.82) and Eq. (2.38) here,

$$f=\frac{\lambda-\kappa}{1+e_0}\ln\frac{p}{p_0}+\frac{\lambda-\kappa}{1+e_0}\ln\left(1+\frac{q^2}{M^2p^2}\right)-\varepsilon_{\mathrm{v}}^{\mathrm{p}}=0 \tag{2.82}$$

$$\frac{\partial f}{\partial\sigma_{ij}}=\frac{\partial f}{\partial p}\frac{\partial p}{\partial\sigma_{ij}}+\frac{\partial f}{\partial q}\frac{\partial q}{\partial\sigma_{ij}} \tag{2.38}$$

where $\partial p/\partial\sigma_{ij}$ and $\partial q/\partial\sigma_{ij}$ are the same as those in the original Cam-clay model. Using Eq. (2.82), we can calculate

$$\begin{aligned}\frac{\partial f}{\partial p}&=\frac{\lambda-\kappa}{1+e_0}\frac{1}{p}+\frac{\lambda-\kappa}{1+e_0}\frac{1}{1+(q^2/(M^2p^2))}\times(-2)\frac{q^2}{M^2}\times\frac{1}{p^3}\\&=\frac{\lambda-\kappa}{1+e_0}\frac{1}{p}\left(1-\frac{2q^2}{M^2p^2+q^2}\right)=\frac{\lambda-\kappa}{1+e_0}\frac{1}{p}\frac{M^2p^2-q^2}{M^2p^2+q^2}\end{aligned} \tag{2.84}$$

$$\frac{\partial f}{\partial q}=\frac{\lambda-\kappa}{1+e_0}\frac{M^2p^2}{M^2p^2+q^2}\times\frac{2q}{M^2p^2}=\frac{\lambda-\kappa}{1+e_0}\frac{2q}{M^2p^2+q^2} \tag{2.85}$$

From Eqs (2.42), (2.48), (2.84), (2.85) and (2.38),

$$\frac{\partial f}{\partial \sigma_{ij}} = \frac{\lambda - \kappa}{1 + e_0}\frac{1}{p}\frac{M^2p^2 - q^2}{M^2p^2 + q^2}\frac{\delta_{ij}}{3} + \frac{\lambda - \kappa}{1 + e_0}\frac{2q}{M^2p^2 + q^2}\frac{3(\sigma_{ij} - p\delta_{ij})}{2q}$$

$$= \frac{\lambda - \kappa}{1 + e_0}\left\{\frac{M^2p^2 - q^2}{M^2p^2 + q^2}\frac{\delta_{ij}}{3p} + \frac{3(\sigma_{ij} - p\delta_{ij})}{M^2p^2 + q^2}\right\} \tag{2.86}$$

Rewriting Eq. (2.31) here,

$$\Lambda = -\frac{(\partial f/\partial p)dp + (\partial f/\partial q)dq}{(\partial f/\partial \varepsilon_{\mathrm{v}}^{\mathrm{p}})(\partial f/\partial \sigma_{ii})} \tag{2.31}$$

and

$$\frac{\partial f}{\partial \varepsilon_{\mathrm{v}}^{\mathrm{p}}} = -1 \tag{2.87}$$

From Eq. (2.86),

$$\frac{\partial f}{\partial \sigma_{ii}} = \frac{\partial f}{\partial \sigma_{ij}}\delta_{ij} = \frac{\lambda - \kappa}{1 + e_0}\frac{1}{p}\frac{M^2p^2 - q^2}{M^2p^2 + q^2} \tag{2.88}$$

From Eqs (2.31), (2.84), (2.85), (2.87), and (2.88), Λ can be calculated as follows:

$$\Lambda = [((\lambda - \kappa)/(1 + e_0))(1/p)((M^2p^2 - q^2)/(M^2p^2 + q^2))dp$$
$$+ (\lambda - \kappa)/(1 + e_0)2q/(M^2p^2 + q^2)dq]$$
$$/[(\lambda - \kappa)/(1 + e_0)(1/p)(M^2p^2 - q^2)/(M^2p^2 + q^2)]$$
$$= dp + \frac{2pq}{M^2p^2 - q^2}dq \tag{2.89}$$

$$\therefore d\varepsilon_{ij}^{\mathrm{p}} = \Lambda\frac{\partial f}{\partial \sigma_{ij}}$$

$$= \frac{\lambda - \kappa}{1 + e_0}\left\{\frac{M^2p^2 - q^2}{M^2p^2 + q^2}\frac{\delta_{ij}}{3p} + \frac{3(\sigma_{ij} - p\delta_{ij})}{M^2p^2 + q^2}\right\}\left(dp + \frac{2pq}{M^2p^2 - q^2}dq\right) \tag{2.90}$$

Therefore, Eq. (2.90) is a general equation for calculating the plastic strain increment $d\varepsilon_{\mathrm{ij}}^{\mathrm{p}}$. As an example, let us discuss the plastic volumetric strain increment $d\varepsilon_{\mathrm{v}}^{\mathrm{p}}$ and the plastic deviatoric strain increment $d\varepsilon_{\mathrm{d}}^{\mathrm{p}}$. From Eq. (2.90), we have

$$d\varepsilon_{11}^{\mathrm{p}} = \frac{\lambda - \kappa}{1 + e_0}\left\{\frac{M^2p^2 - q^2}{M^2p^2 + q^2}\frac{1}{3p} + \frac{3(\sigma_{11} - p)}{M^2p^2 + q^2}\right\}\left(dp + \frac{2pq}{M^2p^2 - q^2}dq\right) \tag{2.91}$$

$$d\varepsilon_{22}^{\mathrm{p}} = \frac{\lambda - \kappa}{1 + e_0}\left\{\frac{M^2p^2 - q^2}{M^2p^2 + q^2}\frac{1}{3p} + \frac{3(\sigma_{22} - p)}{M^2p^2 + q^2}\right\}\left(dp + \frac{2pq}{M^2p^2 - q^2}dq\right) \tag{2.92}$$

$$d\varepsilon_{33}^{\mathrm{p}} = \frac{\lambda - \kappa}{1 + e_0}\left\{\frac{M^2p^2 - q^2}{M^2p^2 + q^2}\frac{1}{3p} + \frac{3(\sigma_{33} - p)}{M^2p^2 + q^2}\right\}\left(dp + \frac{2pq}{M^2p^2 - q^2}dq\right) \tag{2.93}$$

Since $d\varepsilon_v^p = d\varepsilon_{11}^p + d\varepsilon_{22}^p + d\varepsilon_{33}^p$ and $\sigma_{11} + \sigma_{22} + \sigma_{33} = 3p$,

$$d\varepsilon_v^p = \frac{\lambda-\kappa}{1+e_0}\frac{1}{p}\frac{M^2p^2-q^2}{M^2p^2+q^2}\left(dp + \frac{2pq}{M^2p^2-q^2}dq\right) \tag{2.94}$$

$d\varepsilon_v^p$ can be also calculated from Eq. (2.83),

$$\begin{aligned} d\varepsilon_v^p &= \frac{\partial\varepsilon_v^p}{\partial p}dp + \frac{\partial\varepsilon_v^p}{\partial q}dq \\ &= \left(\frac{\lambda-\kappa}{1+e_0}\frac{1}{p} - \frac{\lambda-\kappa}{1+e_0}\frac{M^2p^2}{M^2p^2+q^2}\frac{2q^2}{M^2}\frac{1}{p^3}\right)dp + \frac{\lambda-\kappa}{1+e_0}\frac{M^2p^2}{M^2p^2+q^2}\frac{2q}{M^2p^2}dq \\ &= \frac{\lambda-\kappa}{1+e_0}\frac{1}{p}\frac{M^2p^2-q^2}{M^2p^2+q^2}dp + \frac{\lambda-\kappa}{1+e_0}\frac{2q}{M^2p^2+q^2}dq \end{aligned} \tag{2.95}$$

Here we can easily see that $d\varepsilon_v^p$ calculated by Eq. (2.95) or Eq. (2.94) is the same. From Eq. (2.73), we have

$$\frac{d\varepsilon_v^p}{d\varepsilon_d^p} = \frac{M^2p^2-q^2}{2pq}$$

$d\varepsilon_d^p$ can be directly calculated as follows:

$$d\varepsilon_d^p = \frac{\lambda-\kappa}{1+e_0}\frac{1}{p}\frac{2pq}{M^2p^2+q^2}\left(dp + \frac{2pq}{M^2p^2-q^2}dq\right) \tag{2.96}$$

Therefore, if we express $d\varepsilon_v^p$ and $d\varepsilon_d^p$ in matrix form we have the following simple equation.

$$\begin{Bmatrix} d\varepsilon_v^p \\ d\varepsilon_d^p \end{Bmatrix} = \frac{\lambda-\kappa}{1+e_0}\frac{1}{p}\frac{2pq}{M^2p^2+q^2}\begin{bmatrix} \frac{M^2p^2-q^2}{2pq} & 1 \\ 1 & \frac{2pq}{M^2p^2-q^2} \end{bmatrix}\begin{Bmatrix} dp \\ dq \end{Bmatrix} \tag{2.97}$$

Exercise 2.5 Using the modified Cam-clay model, calculate the stress–strain relations ($q/p - \varepsilon_d$ relation, $\varepsilon_v - \varepsilon_d$ relation, $\sigma_1'/\sigma_3' - \varepsilon_1, \varepsilon_3$ relation, and $\varepsilon_v - \varepsilon_1$ relation) and the effective stress path ($q - p$ relation and $(\sigma_1' - \sigma_3')/2 \sim (\sigma_1' + \sigma_3')/2$ relation) under the following test conditions. Here, σ_1, σ_3, and σ_m are the major, minor, and mean principal stress in terms of total stress, and σ_1', σ_3', and p are the major, minor, and mean principal stress in terms of effective stress while in this book σ_{ij} means the effective stress.

(a) Drained triaxial compression tests and drained triaxial extension tests under $p = 196$ kPa, $\sigma_3' = 196$ kPa, and $\sigma_1' = 196$ kPa
(b) Undrained triaxial compression tests and undrained triaxial extension tests under $\sigma_m = 196$ kPa, $\sigma_3 = 196$ kPa, and $\sigma_1 = 196$ kPa

The values of the model parameters are as follows: $C_c/(1+e_0) = 11.7\%$, $C_s/(1+e_0) = 1.6\%$, $\phi = 34°$, and ν=0.3.

Answer: The method for predicting the stresses-strain relations using the modified Cam-clay model is the same as that using the original Cam-clay model, so see Exercise 2.3 for details. However, different equations should be used in the calculation.
The method for determining the model parameters $(\lambda, \kappa, M, \nu, e_0)$ is also the same as that used in the original Cam-clay model.

(a) Prediction of stress–strain relation under drained condition As an example, Table 2.3 shows the calculated results of the triaxial compression test under $\sigma_3' = 196\,\text{kPa}$. The method for calculating the elastic strain in columns ⑦ and ⑧ is the same as that used in the original Cam-clay model. In calculating the plastic strain increment in columns ⑨ and ⑩, Eqs (2.91) and (2.93) are used.

The solid curves in Figure 2.18 are the predicted results of the triaxial compression test under $\sigma_3' = 196\,\text{kPa}$ from Table 2.3. For reference, the predicted results using the original Cam-clay model are also shown in Figure 2.18 by the dotted curves. It can be seen that with the same model parameters and at the same stress level, the strain predicted by the modified Cam-clay model is smaller than that by the original Cam-clay model. This is because the plastic potential surface of the original Cam-clay model is thinner than that of the modified Cam-clay model. Comparing Figures 2.5 and 2.15, we can see that at the same stress ratio the plastic strain increment vector in the modified Cam-clay model is more inclined to the p-axis than that in the original Cam-clay model. Therefore, the plastic shear strain increment predicted by the modified Cam-clay model is smaller than that predicted by the original Cam-clay model for the same plastic volumetric strain increment.

(b) Prediction of stress–strain relation under undrained condition Substituting the constraint for undrained condition ($d\varepsilon_v = d\varepsilon_v^e + d\varepsilon_v^p = 0$) into Eqs (2.35) and (2.95) gives

$$\frac{\kappa}{1+e_0} \cdot \frac{dp}{p} + \frac{\lambda-\kappa}{1+e_0} \cdot \frac{1}{M^2p^2+q^2} \left\{ \frac{M^2p^2-q^2}{p} dp + 2q\,dq \right\} = 0 \tag{2.98}$$

whcrc

$$p = \sigma_m - u \tag{2.99}$$

$$dp = d\sigma_m - du \tag{2.100}$$

u is the excess pore water pressure.

Here, an undrained triaxial extension test under $\sigma_m = 196\,\text{kPa}$ is taken as an example to explain the prediction of the stress–strain relation and stress path.

Since σ_m = constant,

$$dp = d(\sigma_m - u) = du \tag{2.101}$$

$$dq = d(\sigma_1' - \sigma_3') = d(\sigma_1 - \sigma_3) = d\sigma_1 - d\sigma_3 \tag{2.102}$$

Table 2.3 Predicted results using modified Cam-clay model for drained triaxial compression test at $\sigma'_3 = 196\,\text{kPa}$

① σ'_1 (kPa)	② σ'_3 (kPa)	③ p (kPa)	④ q (kPa)	⑤ dp (kPa)	⑥ dq (kPa)	⑦ $d\varepsilon_1^e$ (%)	⑧ $d\varepsilon_3^e$ (%)	⑨ $d\varepsilon_1^P$ (%)	⑩ $d\varepsilon_3^P$ (%)	⑪ ε_1 (%)	⑫ ε_3 (%)	⑬ ε_d (%)	⑭ ε_v (%)
196.0	196	196.00	0.00							0	0	0	0
245.0	196	212.33	49.00	16.33	49.00	0.134	−0.04	0.327	0.116	0.461	0.076	0.257	0.613
294.0	196	228.67	98.00	16.33	49.00	0.124	−0.037	0.54	0.053	1.125	0.092	0.689	1.308
333.2	196	241.73	137.20	13.07	39.20	0.094	−0.028	0.564	−0.016	1.783	0.048	1.157	1.879
372.4	196	254.80	176.40	13.07	39.20	0.089	−0.027	0.695	−0.084	2.567	−0.062	1.753	2.443
401.8	196	264.60	205.80	9.80	29.40	0.064	−0.019	0.599	−0.106	3.231	−0.187	2.279	2.856
431.2	196	274.40	235.20	9.80	29.40	0.062	−0.019	0.684	−0.154	3.977	−0.36	2.891	3.257
460.6	196	284.20	264.60	9.80	29.40	0.06	−0.018	0.782	−0.21	4.818	−0.588	3.604	3.643
490.0	196	294.00	294.00	9.80	29.40	0.058	−0.017	0.901	−0.276	5.777	−0.881	4.439	4.015
519.4	196	303.80	323.40	9.80	29.40	0.056	−0.017	1.054	−0.36	6.888	−1.258	5.431	4.371
548.8	196	313.60	352.80	9.80	29.40	0.054	−0.016	1.265	−0.472	8.207	−1.747	6.636	4.713
578.2	196	323.40	382.20	9.80	29.40	0.053	−0.016	1.578	−0.636	9.837	−2.398	8.157	5.041
607.6	196	333.20	411.60	9.80	29.40	0.051	−0.015	2.102	−0.904	11.99	−3.318	10.205	5.355
627.2	196	339.73	431.20	6.53	19.60	0.033	−0.01	1.806	−0.808	13.83	−4.136	11.977	5.559
646.8	196	346.27	450.80	6.53	19.60	0.033	−0.01	2.555	−1.185	16.417	−5.33	14.498	5.756
666.4	196	352.80	470.40	6.53	19.60	0.032	−0.01	4.41	−2.115	20.859	−7.455	18.876	5.949
686.0	196	359.33	490.00	6.53	19.60	0.032	−0.009	16.843	−8.334	37.734	−15.799	35.689	6.136

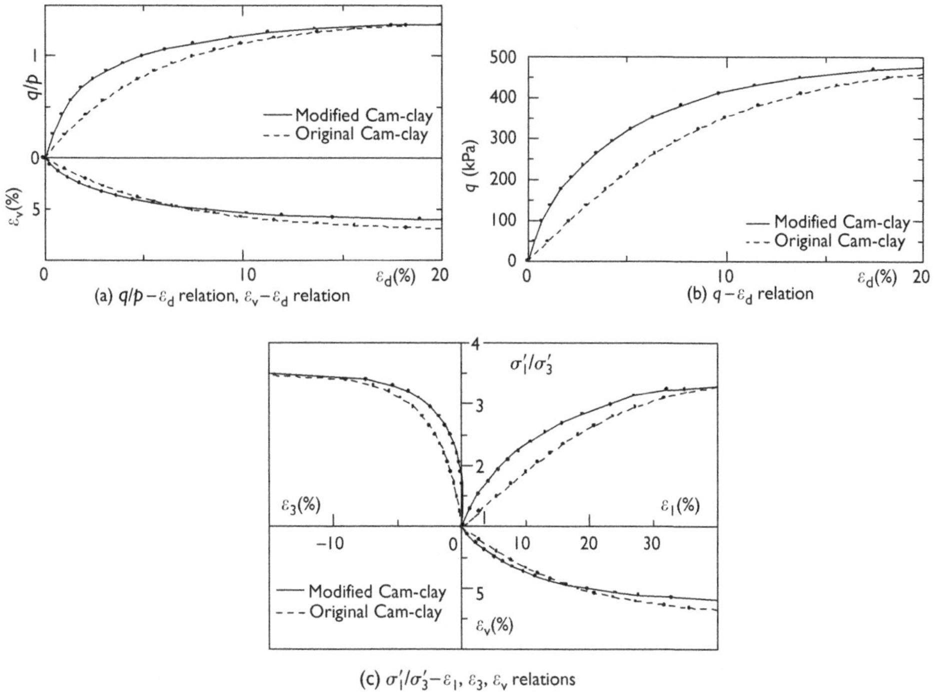

Figure 2.18 Prediction of stress versus strain relation in drained triaxial test by modified Cam-clay model ($\sigma_3' = 196$ kPa).

Substituting Eq. (2.99), (2.101), (2.102) into Eq. (2.98), gives

$$du = \frac{2(\lambda - \kappa)\eta\, dq}{\lambda(M^2 - \eta^2) + 2\kappa\eta^2} \tag{2.103}$$

where,

$$\eta = \frac{q}{\sigma_m - u} \tag{2.104}$$

The procedure for predicting the stress–strain relation under undrained condition using the modified Cam-clay model is the same as that using the original Cam-clay model. Table 2.4 shows the predicted results of the undrained triaxial extension test under $\sigma_m = 196$ kPa. The excess pore water pressures in column ④ are calculated using Eq. (2.103). Other calculation methods are the same as those in drained condition.

The predicted results in Table 2.4 are depicted in Figures 2.19 and 2.20, and it is seen that the effective stress path in the modified Cam-clay model is steeper than that in the original Cam-clay model, and the undrained strength predicted by the modified Cam-clay model is larger than that by the original Cam-clay model. These are due to the following reason. The plastic volumetric strain ε_v^p is the hardening parameter in the Cam-clay models, so the undrained stress path is similar to the yield locus where the plastic strain increment

Table 2.4 Predicted results using modified Cam-clay model for undrained triaxial extension test under $\sigma_m = 196$ kPa

① σ_3 (kPa)	② σ_1 (kPa)	③ σ_m (kPa)	④ du (kPa)	⑤ u (kPa)	⑥ σ_3' (kPa)	⑦ σ_1' (kPa)	⑧ p (kPa)	⑨ q (kPa)	⑩ dp (kPa)	⑪ dq (kPa)	⑫ $d\varepsilon_3^e$ (%)	⑬ $d\varepsilon_1^e$ (%)	⑭ $d\varepsilon_3^p$ (%)	⑮ $d\varepsilon_1^p$ (%)	⑯ ε_3 (%)	⑰ ε_1 (%)	⑱ ε_d (%)	⑲ ε_v (%)
196	196	196		0	196	196	196	0							0	0	0	0
176.4	205.8	196	4.068	4.07	172.33	201.73	191.93	29.4	−4.07	29.4	−0.082	0.034	0.003	0.007	−0.079	0.041	0.08	0.002
166.6	210.7	196	3.153	7.22	159.38	203.48	188.78	44.1	−3.15	14.7	−0.043	0.016	0.001	0.006	−0.121	0.062	0.122	0.004
156.8	215.6	196	4.35	11.57	145.23	204.03	184.43	58.8	−4.35	14.7	−0.045	0.015	0	0.01	−0.167	0.087	0.169	0.006
147	220.5	196	5.706	17.28	129.72	203.22	178.72	73.5	−5.71	14.7	−0.049	0.013	−0.004	0.016	−0.219	0.115	0.223	0.012
137.2	225.4	196	7.32	24.6	112.6	200.8	171.4	88.2	−7.32	14.7	−0.053	0.012	−0.012	0.025	−0.284	0.152	0.291	0.02
127.4	230.3	196	9.37	33.97	93.43	196.33	162.03	102.9	−9.37	14.7	−0.059	0.009	−0.029	0.043	−0.372	0.204	0.384	0.036
119.6	234.2	196	9.427	43.39	76.17	190.83	152.61	114.66	−9.43	11.8	−0.053	0.005	−0.049	0.055	−0.474	0.264	0.492	0.054
117.6	235.2	196	2.686	46.08	71.52	189.12	149.92	117.6	−2.69	2.94	−0.014	0.001	−0.013	0.013	−0.5	0.278	0.519	0.056
113.7	237.2	196	5.994	52.07	61.61	185.09	143.93	123.48	−5.99	5.88	−0.03	0.001	−0.043	0.04	−0.573	0.319	0.595	0.064
109.8	239.1	196	7.013	59.09	50.67	180.03	136.91	129.36	−7.01	5.88	−0.033	−0.001	−0.074	0.061	−0.681	0.379	0.707	0.076
107.8	240.1	196	4.055	63.14	44.66	176.96	132.86	132.3	−4.06	2.94	−0.018	−0.002	−0.048	0.037	−0.747	0.414	0.774	0.08
105.8	241.1	196	4.552	67.7	38.14	173.38	128.3	135.24	−4.55	2.94	−0.02	−0.002	−0.072	0.051	−0.839	0.463	0.868	0.086
103.9	242.1	196	5.23	72.93	30.95	169.13	123.07	138.18	−5.23	2.94	−0.022	−0.004	−0.122	0.08	−0.983	0.538	1.014	0.094
101.9	243	196	6.24	79.16	22.76	163.88	116.84	141.12	−6.24	2.94	−0.025	−0.006	−0.263	0.156	−1.271	0.689	1.307	0.107
100.9	243.5	196	3.844	83.01	17.93	160.52	112.99	142.59	−3.84	1.47	−0.014	−0.005	−0.236	0.132	−1.521	0.816	1.558	0.112
100.5	243.8	196	2.239	85.25	15.2	158.53	110.75	143.32	−2.24	0.74	−0.008	−0.003	−0.187	0.101	−1.716	0.915	1.754	0.113
99.96	244	196	2.506	87.75	12.21	156.27	108.25	144.06	−2.51	0.74	−0.009	−0.004	−0.414	0.216	−2.138	1.127	2.177	0.116
99.76	244.1	196	1.136	88.89	10.87	155.23	107.11	144.35	−1.14	0.29	−0.004	−0.002	−0.293	0.151	−2.436	1.276	2.475	0.116
99.67	244.2	196	0.605	89.5	10.17	154.67	106.5	144.5	−0.61	0.15	−0.002	−0.001	−0.234	0.119	−2.672	1.394	2.711	0.116
99.57	244.2	196	0.629	90.12	9.44	154.09	105.88	144.65	−0.63	0.15	−0.002	−0.001	−0.546	0.275	−3.22	1.668	3.259	0.117
99.53	244.2	196	0.261	90.39	9.14	153.85	105.61	144.71	−0.26	0.06	−0.001	0	−0.451	0.227	−3.672	1.894	3.711	0.117
99.51	244.3	196	0.133	90.52	8.99	153.73	105.48	144.74	−0.13	0.03	0	0	−0.476	0.238	−4.148	2.132	4.187	0.117
99.5	244.3	196	0.08	90.6	8.9	153.65	105.4	144.75	−0.08	0.02	0	0	−0.834	0.417	−4.983	2.55	5.022	0.117
99.49	244.3	196	0.04	90.64	8.85	153.61	105.36	144.76	−0.04	0.01	0	0	−9.484	4.742	−14.47	7.292	14.506	0.117

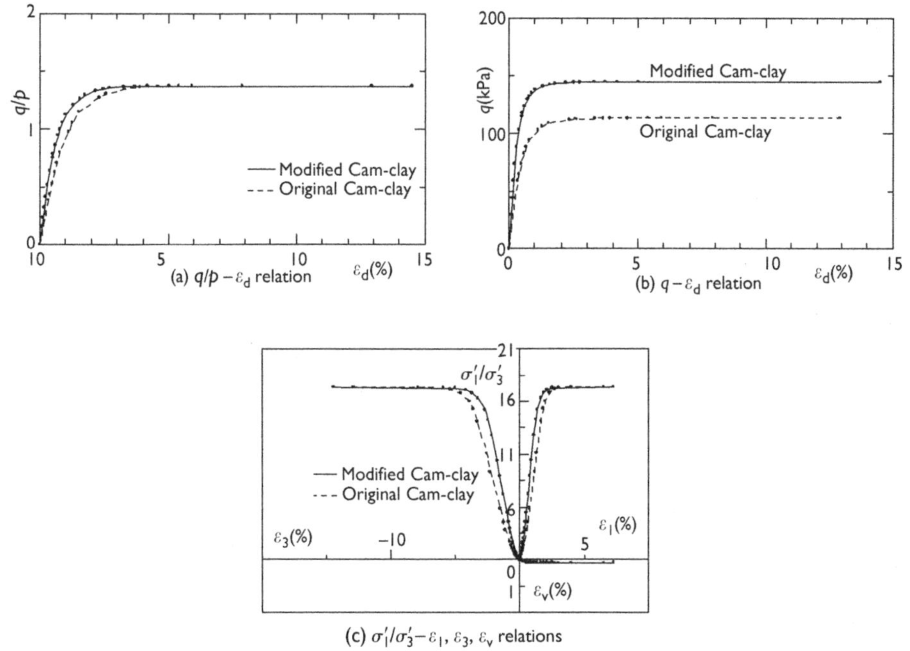

Figure 2.19 Prediction of stress versus strain relation in undrained triaxial extension test by modified Cam-clay model ($\sigma_m = 196$ kPa).

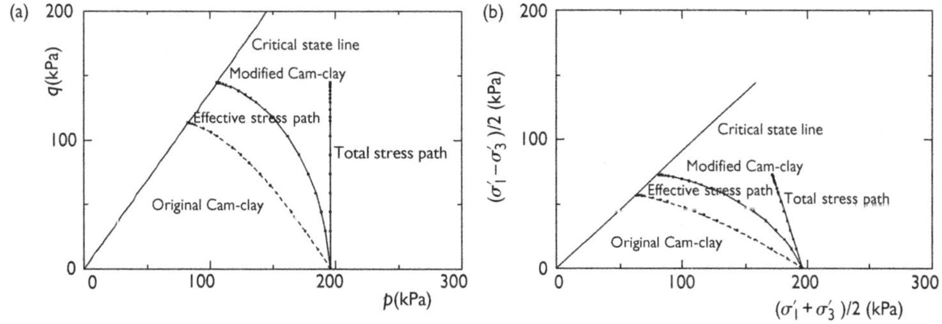

Figure 2.20 Prediction of stress paths in undrained triaxial extension tests by modified Cam-clay model ($\sigma_m = 196$ kPa).

$d\varepsilon_v^p = 0$ because the elastic volumetric increment $d\varepsilon_v^e$ is very small in general. As shown in Figure 2.16, the yield locus of the modified Cam-clay model is steeper, so the intersection between the critical state line and the yield locus is higher.

There is a very small volumetric strain during shearing in Figure 2.19(c). The reason is the same as explained for Figure 2.12(c).

References

Hata, S., Ohta, H., and Yoshitami, S. 1969. On the state surface of soils. *Proceedings of JSCE*, 172: 97–117.

Henkel, D.J. 1960. The relationships between the effective stress and water content in saturated clays. *Geotechnique*, 10(2): 41–54.

Newland, P.L. and Allely, B.H. 1957. Volume change in drained triaxial tests on granular materials. *Geotechnique*, 7(1): 17–34.

Ohta, H. 1993. Application of flow rule and Cam-clay model. *Tsuchi-to-Kiso*, 41(9): 61–68.

Roscoe, K.H. and Burland, J.B. 1968. *On the Generalised Stress-strain Behaviour of 'Wet' Clay.* Engineering Plasticity, Cambridge University Press, Cambridge, UK, pp. 535–609.

Roscoe, K.H., Schofield, A.N., and Thurairajah, A. 1963. Yielding of clay in states wetter than critical. *Geotechnique*, 13(3): 211–240.

Rowe, P.W. 1962. The stress-dilatancy relation for static equilibrium of an assembly of particle in contact. *Proceedings of the Royal Society* A. 269: 500–527.

Schofield, A.N. and Wroth, C.P. 1968. *Critical State Soil Mechanics*, McGraw-Hill, New York.

Tatsuoka, F. 1978. Fundamental study on deformation of granular materials; part 1. *Tsuchi-to-Kiso*, 27(6): 82–89.

Wood, D.M. 1990. *Soil Behaviour and Critical State Soil Mechanics*. Cambridge University Press.

Chapter 3

The Cam-clay model revised by the SMP criterion

3.1 Introduction

The Cam-clay models are popular and fundamental constitutive models for normally consolidated clay. As described in Chapter 2, the original Cam-clay model was proposed by Roscoe *et al.* (1963) and the modified Cam-clay model was introduced by Roscoe and Burland (1968). The stress invariants used in the Cam-clay models are p and q, which are expressed in terms of the principal stress σ_i or the stress tensor σ_{ij} as follows

$$\left.\begin{aligned} p &= \tfrac{1}{3}(\sigma_1+\sigma_2+\sigma_3) = \tfrac{1}{3}\sigma_{ii} \\ q &= \tfrac{1}{\sqrt{2}}\sqrt{(\sigma_1-\sigma_2)^2+(\sigma_2-\sigma_3)^2+(\sigma_3-\sigma_1)^2} = \sqrt{\tfrac{3}{2}(\sigma_{ij}-p\delta_{ij})(\sigma_{ij}-p\delta_{ij})} \end{aligned}\right\} \tag{3.1}$$

where δ_{ij} is the Kronecker delta. It is seen from Eq. (3.1) that $p = \sigma_{\text{oct}}$ (σ_{oct}: normal stress on the octahedral plane) and $q = (3/\sqrt{2})\tau_{\text{oct}}$ (τ_{oct}: shear stress on the octahedral plane). Therefore, the Extended Mises criterion ($q/p = \text{const}$ or $\tau_{\text{oct}}/\sigma_{\text{oct}} = \text{const}$) was adopted for shear yield and shear failure of clay in the Cam-clay models. The shear yield is caused by increase of the stress ratio $\eta(= q/p)$, while the compressive yield is caused by increase of the mean stress p.

However, it is well known that the failure of soil cannot be properly described by the Extended Mises criterion, but rather by criteria such as the Mohr-Coulomb criterion or the SMP criterion (Matsuoka and Nakai 1974; Matsuoka 1976). Taking into account the consistency in shear deformation and shear failure, it is quite natural to introduce the Mohr-Coulomb criterion or the SMP criterion for shear yield as well as for shear failure of soils. Based on this rationale, some researchers attempted to extend the Cam-clay models to three-dimensional models for soils (e.g. Nayak and Zienkiewicz 1972; Zienkiewicz and Pande 1977; Van Eekelen 1980; Randolph 1982; Nakai and Mihara 1984). For example, the so-called $g(\theta)$ method, which has widely been used in numerical implementation of the Cam-clay models, considers the variation of the yield surface with the Lode angle θ. However, such a method may results in a non-convex yield surface when the friction angle is large (Sheng *et al.* 2000). As pointed out by Wroth and Houlsby (1985), who is one of the developers of the Cam-clay models, further study is necessary to improve the Cam-clay models by combining the critical state theory with failure criteria such as those by Matsuoka and Nakai (1974) and Lade and Duncan (1975).

In this chapter, a method for the transformation of the curved surface of the SMP criterion to a cone in the transformed principal stress space is proposed by introducing a transformed stress $\tilde{\sigma}_{ij}$ (Matsuoka, Yao, and Sun 1999). The transformed stress $\tilde{\sigma}_{ij}$ is applied to the original and modified Cam-clay models, which are then called the revised Cam-clay models by the SMP criterion. Compared to the Cam-clay models, the revised models have the following main features:

1 using $\tilde{\sigma}_{ij}$ instead of σ_{ij},
2 using the same set of strain variables as the Cam-clay models (i.e. ε_{v} and ε_{d}),
3 using the same set of model parameters as the Cam-clay models (i.e. M, λ, κ, e_0, and ν)
4 recovering the Cam-clay models under triaxial compression
5 with the shape of the yield surface in the deviatoric plane being similar to the SMP failure criterion.

These revised models are a combination of the Cam-clay models and the SMP criterion and have realized the consistency from shear yield to shear failure of clay, both of which obey the SMP criterion.

This chapter will first describe the transformed stress $\tilde{\sigma}_{ij}$ and the revised Cam-clay models. The model prediction will then be compared with the measured data for drained and undrained tests of clay under triaxial compression, triaxial extension, plane strain, and true triaxial conditions. The comparisons indicate that the Cam-clay models using the transformed stress tensor $\tilde{\sigma}_{ij}$ can describe well the drained and undrained behavior of clay under triaxial compression, but also under triaxial extension, plane strain, and true triaxial conditions. Therefore, the stress transformation method provides a reasonable and simple approach for incorporating the three-dimensional SMP criterion into existing models based on two stress invariants p and q.

Throughout this chapter, the term "stress" is to be interpreted as effective stress, the term "Cam-clay models" as the original Cam-clay model and the modified Cam-clay model, and the term "revised Cam-clay models" as the original Cam-clay model revised by the SMP criterion and the modified Cam-clay model revised by the SMP criterion.

3.2 The SMP criterion and a transformed stress tensor

Figures 3.1 and 3.2 show the shapes of the Extended Tresca, the Extended Mises, the Mohr-Coulomb, and the SMP failure criteria in the principal stress space and in the

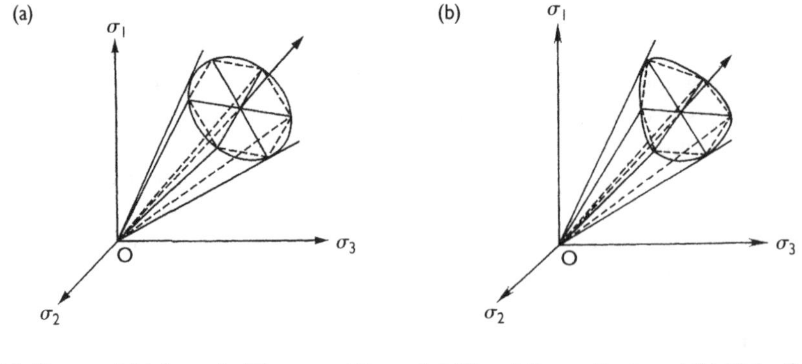

Figure 3.1 Shapes of (a) Extended Tresca and Extended Mises failure criteria and (b) Mohr-Coulomb and SMP criteria in principal stress space.

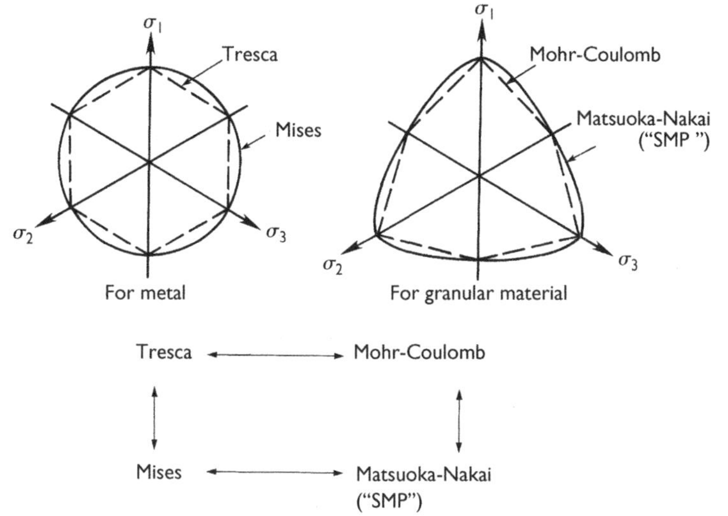

Figure 3.2 Mutual relationships between Extended Tresca and Extended Mises failure criteria and Mohr-Coulomb and SMP criteria in π-plane.

π-plane, respectively. We can see from Figure 3.1 that, just as the cone of the Extended Mises criterion circumscribes the regular hexagonal pyramid of the Extended Tresca criterion, the SMP criterion is a smoothly curved convex surface circumscribing the irregular hexagonal pyramid of the Mohr-Coulomb criterion in principal stress space. We can also see from Figure 3.2 that, just as the circle of the Extended Mises criterion circumscribes the regular hexagon of the Extended Tresca criterion, the SMP criterion is a smooth and convex curve circumscribing the irregular hexagon of the Mohr-Coulomb criterion in the π-plane.

Figure 3.3 shows the Extended Mises, the Mohr-Coulomb, and the SMP criteria compared with the experimental failure stress states obtained from triaxial compression, triaxial extension, and true triaxial tests on Fujinomori clay (Nakai and Matsuoka 1986). It is clear that the experimental data (marked as ○ and ■) are closest to the SMP failure criterion represented by the solid curve and are furthest from the Extended Mises failure criterion represented by the dash-dotted curve. The latter is unfortunately adopted in the Cam-clay models.

Realizing that the Extended Mises failure criterion is not so appropriate for clay, some researchers have attempted to adopt the Mohr-Coulomb failure criterion in their models. However, they continue to use p and q as the stress variables alone, which means that shear yielding of soil obeys the Extended Mises criterion. Considering the consistency and continuation in shear deformation and shear failure, it is reasonable to assume that the Mohr-Coulomb or the SMP criterion is valid not only for shear failure, but also for shear yield. As it can be seen from Figure 3.3 that the prediction of the SMP failure criterion is better than that of the Mohr-Coulomb failure criterion, we select the SMP criterion rather than the Mohr-Coulomb criterion as the criterion for shear failure and shear yield for the revised Cam-clay models.

In order to combine the SMP criterion with the Cam-clay model, a transformed stress $\tilde{\sigma}_{ij}$ is proposed (Matsuoka, Yao, and Sun 1999). The curved surface of the SMP criterion in

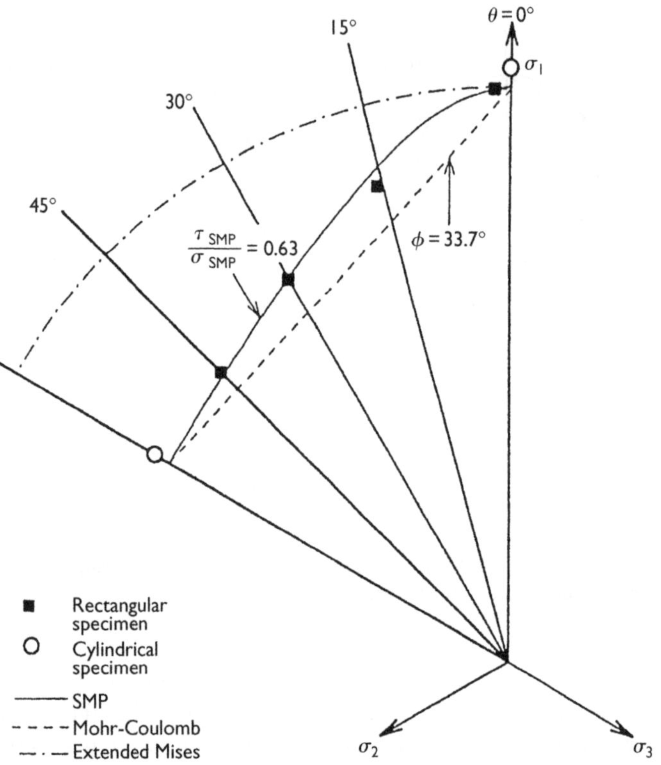

Figure 3.3 Comparison of Extended Mises, Mohr-Coulomb, and SMP criteria with stress states at failure in π-plane obtained by triaxial compression, triaxial extension, and true triaxial tests on Fujinomori clay (after Nakai and Matsuoka 1986).

the principal stress space, as shown in Figure 3.1(b), can be transformed into a cone in the transformed principal stress space. So, we can use the Cam-clay model in the transformed stress space instead of the conventional stress space. How to obtain the transformed stress tensor $\tilde{\sigma}_{ij}$ is described as follows.

When the value of τ_{SMP}/σ_{SMP} is given, the shape and size of the SMP criterion in principal stress space (Fig. 3.1(b)) is defined. The length of OC shown in Figure 3.4 for a given p can be expressed in the following function.

$$\ell_0 = f(\tau_{SMP}/\sigma_{SMP}, p) \tag{3.2}$$

The particular expression of ℓ_0 is given as follows (see Appendix 1 for the details):

$$\ell_0 = \sqrt{\frac{2}{3}}\frac{6p}{3\sqrt{1+8(\tau_{SMP}/\sigma_{SMP})^{-2}/9}-1} = \sqrt{\frac{2}{3}}\frac{2I_1}{3\sqrt{(I_1I_2-I_3)/(I_1I_2-9I_3)}-1} \tag{3.3}$$

In order to transform the curve of the SMP criterion with OC $= \ell_0$ in the π-plane of σ_i space (Fig. 3.4) into a circle with a radius of ℓ_0 in the π-plane of $\tilde{\sigma}_i$ (the principal value of

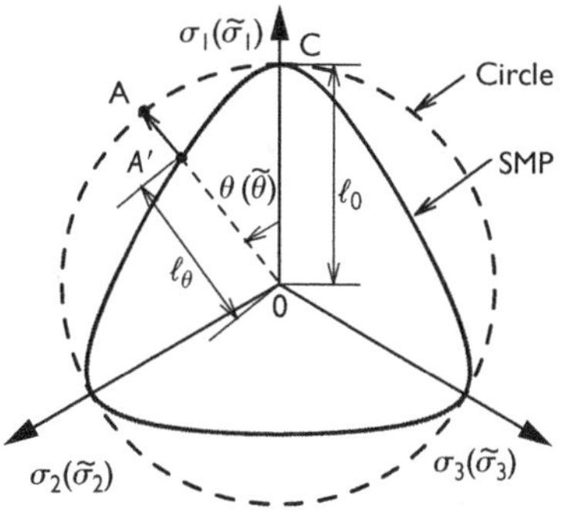

Figure 3.4 The SMP criterion in π-plane (solid line) and transformed π-plane (broken circle).

$\tilde{\sigma}_{ij}$) space (Fig. 3.4) under the same angle ($\tilde{\theta} = \theta$) between the $\tilde{\sigma}_1$ or σ_1 axis and the direction of the transformed stress vector or the direction of the ordinary stress vector in the π-plane (Fig. 3.4), and under the same mean value of $\tilde{\sigma}_i$ and σ_i, the following equations can be made:

$$\tilde{p} = p \tag{3.4}$$

$$\tilde{\theta} = \theta \tag{3.5}$$

$$\sqrt{\tilde{s}_{ij}\tilde{s}_{ij}} = \ell_0 \tag{3.6}$$

where θ and $\tilde{\theta}$ are Lode's angles of the ordinary and the transformed stresses (Fig. 3.4), i.e.

$$\theta = \frac{1}{3}\arccos\left(\sqrt{6}\frac{s_{ik}s_{kl}s_{li}}{(s_{mn}s_{mn})^{3/2}}\right) \tag{3.7}$$

$$\tilde{\theta} = \frac{1}{3}\arccos\left(\sqrt{6}\frac{\tilde{s}_{ik}\tilde{s}_{kl}\tilde{s}_{li}}{(\tilde{s}_{mn}\tilde{s}_{mn})^{3/2}}\right) \tag{3.8}$$

$\tilde{s}_{ij}$ is the transformed deviatoric stress tensor ($= \tilde{\sigma}_{ij} - \tilde{p}\delta_{ij}$) and $\tilde{p}$ is the transformed mean stress, i.e.,

$$\tilde{p} = \tfrac{1}{3}(\tilde{\sigma}_1 + \tilde{\sigma}_2 + \tilde{\sigma}_3) = \tfrac{1}{3}\tilde{\sigma}_{ii} \tag{3.9}$$

Taking Eqs (3.7) and (3.8) into consideration, Eqs (3.5), (3.6) can be rewritten as

$$\frac{\tilde{s}_{ik}\tilde{s}_{kl}\tilde{s}_{li}}{(\tilde{s}_{mn}\tilde{s}_{mn})^{3/2}} = \frac{s_{ik}s_{kl}s_{li}}{(s_{mn}s_{mn})^{3/2}} \tag{3.10}$$

$$\sqrt{\tilde{s}_{ij}\tilde{s}_{ij}} = \frac{\ell_0}{\ell_\theta}\sqrt{s_{ij}s_{ij}} \tag{3.11}$$

where ℓ_θ is the length of OA′, as shown in Figure 3.4.

In the condition that the principal directions of $\tilde{\sigma}_{ij}$ and σ_{ij} are the same, as shown in Figure 3.4, $\tilde{\sigma}_{ij}$ of point A, which corresponds to σ_{ij} of point A′, can be obtained from

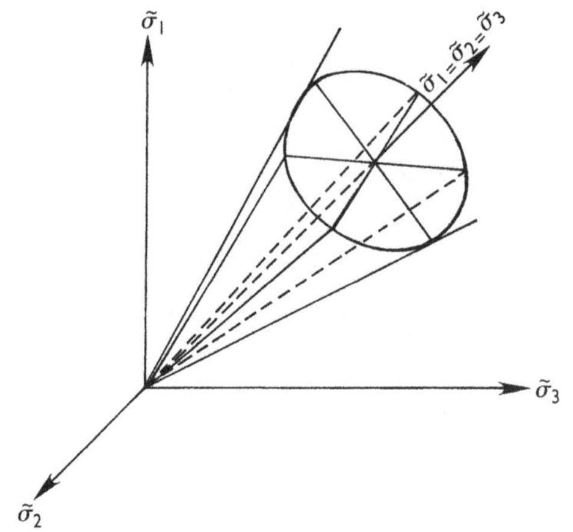

Figure 3.5 Shapes of SMP failure criterion expressed in transformed principal stress space.

Eqs (3.4), (3.10), and (3.11) as follows:

$$\tilde{\sigma}_{ij} = \tilde{p}\delta_{ij} + \tilde{s}_{ij} = \tilde{p}\delta_{ij} + \frac{\ell_0}{\ell_\theta} s_{ij} = p\delta_{ij} + \frac{\ell_0}{\sqrt{s_{kl}s_{kl}}} s_{ij} \tag{3.12}$$

When the stress σ_{ij} is given, ℓ_0 can be calculated from Eq. (3.3), and then the transformed stress $\tilde{\sigma}_{ij}$ can be calculated from Eq. (3.12).

From this derivation, it is known that the shape of the SMP criterion becomes a cone with the axis being the space diagonal $\tilde{\sigma}_1 = \tilde{\sigma}_2 = \tilde{\sigma}_3$ in the transformed principal stress space (see Fig. 3.5), or becomes a circle with the center being the origin O in the $\tilde{\pi}$-plane (see Fig. 3.4). Taking into consideration the similarity in the shapes of the Extended Mises criterion in the principal stress space and the SMP criterion in the transformed principal stress space, i.e. both are cones with the axis being the space diagonal, we can revise the Cam-clay models by using the transformed stress tensor $\tilde{\sigma}_{ij}$.

3.3 The Cam-clay model revised by the SMP criterion

In this part, we present the two Cam-clay models (the original and the modified) revised by using the SMP criterion and the transformed stress. The difference between the revised Cam-clay models and the Cam-clay models is only the stress tensor used in the models. In the revised Cam-clay models, the transformed stress tensor $\tilde{\sigma}_{ij}$ is used to replace the stress tensor σ_{ij}. Table 3.1 compares the original Cam-clay model with the revised original Cam-clay model by the SMP criterion. It can be seen that the stress and strain variables, the failure criteria, the stress–dilatancy equations, the plastic potential and yield functions, the hardening parameters, and the material parameters in the two models are all the same, except that the stress tensor σ_{ij} exchanges with $\tilde{\sigma}_{ij}$.

Let's discuss the critical state conditions, the shapes of the stress–dilatancy equation and the yield function of the two models under triaxial compression and triaxial extension stress

Table 3.1 Comparison of original Cam-clay model with revised original Cam-clay model by the SMP criterion

<table>
<tr><th>Content</th><th>Original Cam-clay model</th><th>Revised original Cam-clay model</th></tr>
<tr><td>Stress tensor</td><td>σ_{ij}</td><td>$\tilde{\sigma}_{ij} = p\delta_{ij} + \frac{\ell_0}{\ell_\theta}(\sigma_{ij} - p\delta_{ij})$</td></tr>
<tr><td>Stress invariant</td><td>$p = \frac{1}{3}\sigma_{ii}, q = \sqrt{\frac{3}{2}(\sigma_{ij} - p\delta_{ij})(\sigma_{ij} - p\delta_{ij})}$</td><td>$\tilde{p} = \frac{1}{3}\tilde{\sigma}_{ii}, \tilde{q} = \sqrt{\frac{3}{2}(\tilde{\sigma}_{ij} - \tilde{p}\delta_{ij})(\tilde{\sigma}_{ij} - \tilde{p}\delta_{ij})}$</td></tr>
<tr><td>Strain increment</td><td colspan="2">$d\varepsilon_{ij}$</td></tr>
<tr><td>Strain increment invariant</td><td colspan="2">$d\varepsilon_v = d\varepsilon_{ii}, d\varepsilon_d = \sqrt{\frac{2}{3}\left(d\varepsilon_{ij} - \frac{1}{3}d\varepsilon_v\delta_{ij}\right)\left(d\varepsilon_{ij} - \frac{1}{3}d\varepsilon_v\delta_{ij}\right)}$</td></tr>
<tr><td>Total strain increment</td><td colspan="2">$d\varepsilon_{ij} = d\varepsilon_{ij}^e + d\varepsilon_{ij}^p$</td></tr>
<tr><td>Elastic strain increment</td><td colspan="2">$d\varepsilon_{ij}^e = \frac{1+\nu}{E}d\sigma_{ij} - \frac{\nu}{E}d\sigma_{mm}\delta_{ij}, E = \frac{3(1-2\nu)(1+e_0)}{\kappa}p$</td></tr>
<tr><td>Failure criterion</td><td>$q/p = M$</td><td>$\tilde{q}/\tilde{p} = M$</td></tr>
<tr><td></td><td colspan="2">$M = 6\sin\phi/(3 - \sin\phi)$</td></tr>
<tr><td>Stress–dilatancy equation</td><td>$\frac{q}{p} = M - \frac{d\varepsilon_v^p}{d\varepsilon_d^p}$</td><td>$\frac{\tilde{q}}{\tilde{p}} = M - \frac{d\varepsilon_v^p}{d\varepsilon_d^p}$</td></tr>
<tr><td>Normality condition</td><td>$\frac{dq}{dp} \times \frac{d\varepsilon_d^p}{d\varepsilon_v^p} = -1$</td><td>$\frac{d\tilde{q}}{d\tilde{p}} \times \frac{d\varepsilon_d^p}{d\varepsilon_v^p} = -1$</td></tr>
<tr><td>Plastic potential and yield function</td><td>$f = g = \frac{\lambda - \kappa}{1 + e_0}\left[\frac{q}{Mp} + \ln\frac{p}{p_0}\right] - \varepsilon_v^p = 0$</td><td>$f = g = \frac{\lambda - \kappa}{1 + e_0}\left[\frac{\tilde{q}}{M\tilde{p}} + \ln\frac{\tilde{p}}{\tilde{p}_0}\right] - \varepsilon_v^p = 0$</td></tr>
<tr><td>Hardening parameter</td><td colspan="2">ε_v^p</td></tr>
<tr><td>Plastic strain increment (Flow rule)</td><td>$d\varepsilon_{ij}^p = \Lambda\frac{\partial f}{\partial \sigma_{ij}}$</td><td>$d\varepsilon_{ij}^p = \Lambda\frac{\partial f}{\partial \tilde{\sigma}_{ij}}$</td></tr>
<tr><td>Proportionality constant</td><td>$\Lambda = dp + \frac{dq}{M - q/p}$</td><td>$\Lambda = d\tilde{p} + \frac{d\tilde{q}}{M - \tilde{q}/\tilde{p}}$</td></tr>
<tr><td>Stress gradient</td><td>$\frac{\partial f}{\partial \sigma_{ij}} = \frac{\lambda - \kappa}{1 + e_0}\frac{1}{Mp}\left[\frac{1}{3}\left(M - \frac{q}{p}\right)\delta_{ij} + \frac{3(\sigma_{ij} - p\delta_{ij})}{2q}\right]$</td><td>$\frac{\partial f}{\partial \tilde{\sigma}_{ij}} = \frac{\lambda - \kappa}{1 + e_0}\frac{1}{M\tilde{p}}\left[\frac{1}{3}\left(M - \frac{\tilde{q}}{\tilde{p}}\right)\delta_{ij} + \frac{3(\tilde{\sigma}_{ij} - \tilde{p}\delta_{ij})}{2\tilde{q}}\right]$</td></tr>
<tr><td>Model parameter</td><td colspan="2">$\phi, \lambda/(1 + e_0), \kappa/(1 + e_0), \nu$</td></tr>
</table>

Notes
e_0: initial void ratio; ϕ: angle of internal friction; λ: compression index; κ: swelling index; *E*: elastic modulus; ν: Poisson's ratio; ℓ_0: seeing Eq. (3.3) and Figure 3.4; ℓ_θ: see Eq. (3.12) and Figure 3.4.

states. From Table 3.1, the critical state conditions of the revised original Cam-clay model under three-dimensional stress states can be expressed as follows:

$$\tilde{q}_{cs}/\tilde{p}_{cs} = M \tag{3.13}$$

$$\varepsilon_v^p = \frac{\lambda - \kappa}{1 + e_0}\left(\ln \frac{\tilde{p}_{cs}}{\tilde{p}_0} + 1\right) \tag{3.14}$$

where e_0 is the void ratio for $\tilde{p} = \tilde{p}_0$, suffix cs stands for critical state. When Eqs (3.13) and (3.14) are satisfied, the soil continues to deform under constant stresses and constant volume. Equations (3.13) and (3.14) are the same as the critical state conditions of the original Cam-clay model under triaxial compression and Eq. (3.13) satisfies the SMP failure criterion, so we may say Eqs (3.13) and (3.14) are the extensions for the critical state condition of the Cam-clay model under three-dimensional stress states.

The stress–dilatancy equation of the revised original Cam-clay model, $\tilde{q}/\tilde{p} = M - d\varepsilon_v^p/d\varepsilon_d^p$, can be depicted as Figure 3.6(a) in the $\tilde{q}/\tilde{p} \sim -d\varepsilon_v^p/d\varepsilon_d^p$ plane or as Figure 3.6(b) in the $q/p \sim -d\varepsilon_v^p/d\varepsilon_d^p$ plane. The line in Figure 3.6(a) is applicable to triaxial compression and triaxial extension, as well as other stress states. This line becomes two lines in the $q/p \sim -d\varepsilon_v^p/d\varepsilon_d^p$ plane under triaxial compression and triaxial extension (Fig. 3.6(b)). Figure 3.7 shows the results of triaxial compression and triaxial extension tests on Fujinomori clay, plotted in the $\tilde{q}/\tilde{p} \sim -d\varepsilon_v^p/d\varepsilon_d^p$ plane (Fig. 3.7(a)) and in the $q/p \sim -d\varepsilon_v^p/d\varepsilon_d^p$ plane (Fig. 3.7(b)). It can be seen from these two figures that the adopted stress–dilatancy equation in the revised original Cam-clay model can well describe the observed test results. That is to say, the unique relationship between $\tilde{q}/\tilde{p}$ and $-d\varepsilon_v^p/d\varepsilon_d^p$ (Fig. 3.6(a)) can describe the observed difference of the $q/p \sim -d\varepsilon_v^p/d\varepsilon_d^p$ curves between triaxial compression and triaxial extension (Fig. 3.7(b)). However, the stress–dilatancy equation of $q/p = M - d\varepsilon_v^p/d\varepsilon_d^p$ used in the original Cam-clay model cannot describe this difference (Schofield and Wroth 1968).

Under triaxial compression and triaxial extension conditions, the yield function in Table 3.1 is plotted in Figure 3.8. The transformed stresses $\tilde{\sigma}_a$ and $\tilde{\sigma}_r$ correspond to σ_a and σ_r respectively, with σ_a and σ_r being the axial and radial stresses. It can be seen from Figure 3.8 that, although the yield loci under triaxial compression and triaxial extension are symmetrical about $\tilde{p}$ axis in the $\tilde{p} \sim (\tilde{\sigma}_a - \tilde{\sigma}_r)$ plane (see Fig. 3.8(a)), they are not symmetrical with respect to p axis in the $p \sim (\sigma_a - \sigma_r)$ plane. The value of q in triaxial extension is less than

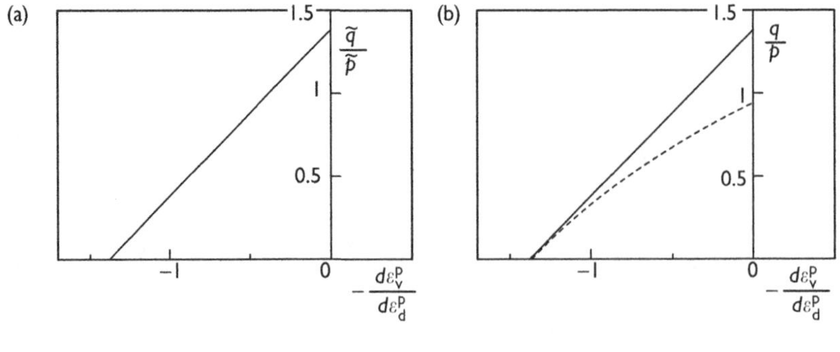

Figure 3.6 Stress–dilatancy relationships of the revised original Cam-clay model under triaxial compression and triaxial extension.

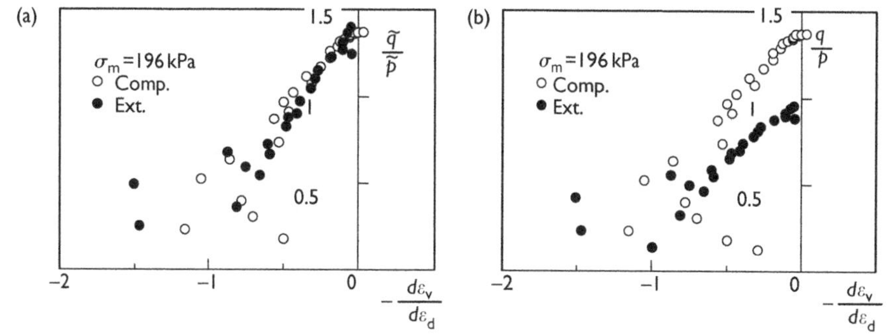

Figure 3.7 Stress–dilatancy relationships for Fujinomori clay under triaxial compression and triaxial extension.

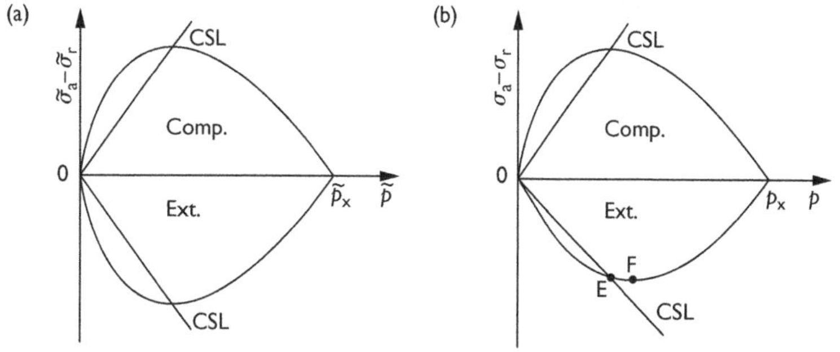

Figure 3.8 Yield loci of original Cam-clay model generalized by the SMP criterion in triaxial compression and triaxial extension.

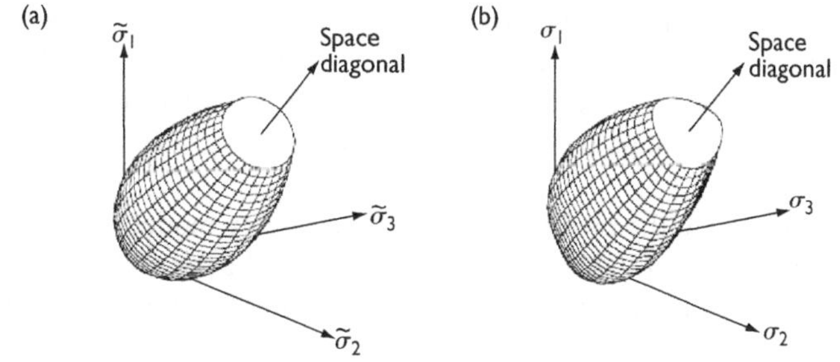

Figure 3.9 Yield surface expressed in (a) transformed principal stress space and (b) ordinary principal stress space.

that in triaxial compression in the $p \sim (\sigma_a - \sigma_r)$ plane at the same p (see Fig. 3.8(b)). This trend is similar to the test results for various kinds of soils (e.g. Mitachi and Kitago 1979; Miura *et al.* 1984). Figure 3.9 shows the yield surface of the revised Cam-clay model in the transformed principal stress space and in the ordinary principal stress space. It can be seen from Figure 3.9(a) that the shape of the yield surface of the revised original Cam-clay model in the transformed principal stress space is the same as that of the original Cam-clay model in the ordinary principal stress space (i.e. a surface resulting from the rotating the two-dimensional curve in the meridian plane about the space diagonal). On the other hand, we can see from Figure 3.9(b) that the yield surface in the ordinary principal stress space is not a result of the rotation of the two-dimensional curve and the yield locus in the π-plane is similar to the SMP curve.

Table 3.2 compares the modified Cam-clay model with the revised modified Cam-clay model by the SMP criterion. The stress and strain increment variables used in these models are identical to their counterparts in Table 3.1, and are thus not given in this table.

The stress–dilatancy equation adopted in the revised modified Cam-clay model $\left(d\varepsilon_{\mathrm{v}}^{\mathrm{p}}/d\varepsilon_{\mathrm{d}}^{\mathrm{p}} = (M^2\tilde{p}^2 - \tilde{q}^2)/(2\tilde{p}\tilde{q})\right)$ can be depicted as a unique curve shown in Figure 3.10(a) in the $\tilde{q}/\tilde{p} \sim -d\varepsilon_{\mathrm{v}}^{\mathrm{p}}/d\varepsilon_{\mathrm{d}}^{\mathrm{p}}$ plane and as two curves shown in Figure 3.10(b) in the $q/p \sim -d\varepsilon_{\mathrm{v}}^{\mathrm{p}}/d\varepsilon_{\mathrm{d}}^{\mathrm{p}}$ plane for triaxial compression and triaxial extension. The yield function in Table 3.2 under triaxial compression and triaxial extension is also drawn in Figure 3.11. These features of the revised model are again better than those of the modified Cam-clay model in describing the stress path dependency of the stress–dilatancy equation and in describing the yield function for soils.

As shown in Tables 3.1 and 3.2, the associated flow rule is adopted in the transformed stress $\tilde{\sigma}_{ij}$ space in the two revised models, i.e.

$$d\varepsilon_{ij}^{\mathrm{p}} = \Lambda \frac{\partial f}{\partial \tilde{\sigma}_{ij}} \tag{3.15}$$

This associativity in the transformed stress space actually means the loss of the associativity in the ordinary stress σ_{ij} space, excluding triaxial compression stress condition ($\tilde{\sigma}_{ij} = \sigma_{ij}$). One of the reasons why an associated flow rule is not adopted in the ordinary stress σ_{ij} space is given in the following passage.

The maximum deviatoric stress point F on the yield locus in triaxial extension is located on the right side of the CSL, as shown in Figures 3.8(b) and 3.11(b). This leads to the contradiction that the negative plastic volumetric strain would occur in the region EF if the associated flow rule was adopted and the hardening parameter was the plastic volumetric strain. In order to avoid this contradiction and to use the SMP criterion in the revised models, the associated flow rule is adopted in the transformed stress $\tilde{\sigma}_{ij}$ space instead of the ordinary stress σ_{ij} space.

In addition, compared to the modified stress t_{ij} and t_{ij}-clay model, which was also developed from the SMP concept (Nakai and Matsuoka 1986), the transformed stress $\tilde{\sigma}_{ij}$ and the revised models described in this chapter have the following features:

1 In the transformed π-plane, the shape of the SMP criterion is a circle (Fig. 3.4), and in the transformed principal stress space, the shape of the SMP criterion is a cone. On the other hand, the SMP criterion is not a circle in the π-plane of t_{ij} space and is not a cone in the principal space of t_{ij}.
2 Any other elastoplastic models for soils, which are checked only in triaxial compression, can be extended simply and directly to three-dimensional stress conditions by using the transformed stress $\tilde{\sigma}_{ij}$ instead of the ordinary stress σ_{ij}. On the other hand, the modified stress t_{ij} does not have this function.

Table 3.2 Comparison of modified Cam-clay model with revised modified Cam-clay model by the SMP criterion

Content	*Modified Cam-clay model*	*Revised modified Cam-clay model*
Failure criterion	$q/p = M$	$\tilde{q}/\tilde{p} = M$
	$M = 6\sin\phi/(3 - \sin\phi)$	
Stress–dilatancy equation	$\dfrac{d\varepsilon_v^P}{d\varepsilon_d^P} = \dfrac{M^2p^2 - q^2}{2pq}$	$\dfrac{d\varepsilon_v^P}{d\varepsilon_d^P} = \dfrac{M^2\tilde{p}^2 - \tilde{q}^2}{2\tilde{p}\tilde{q}}$
Normality condition	$\dfrac{dq}{dp} \times \dfrac{d\varepsilon_d^P}{d\varepsilon_v^P} = -1$	$\dfrac{d\tilde{q}}{d\tilde{p}} \times \dfrac{d\varepsilon_d^P}{d\varepsilon_v^P} = -1$
Plastic potential and yield function	$f = \psi = \dfrac{\lambda - \kappa}{1 + e_0}\left[\ln\dfrac{p}{p_0} + \ln\dfrac{M^2p^2 + q^2}{M^2p^2}\right] - \varepsilon_v^P = 0$	$f = \psi = \dfrac{\lambda - \kappa}{1 + e_0}\left[\ln\dfrac{\tilde{p}}{\tilde{p}_0} + \ln\dfrac{M^2\tilde{p}^2 + \tilde{q}^2}{M^2\tilde{p}^2}\right] - \varepsilon_v^P = 0$
Hardening parameter	ε_v^P	
Plastic strain increment (Flow rule)	$d\varepsilon_{ij}^P = \Lambda\dfrac{\partial f}{\partial\sigma_{ij}}$	$d\varepsilon_{ij}^P = \Lambda\dfrac{\partial f}{\partial\tilde{\sigma}_{ij}}$
Proportionality constant	$\Lambda = dp + \dfrac{2pq}{M^2p^2 - q^2}dq$	$\Lambda = d\tilde{p} + \dfrac{2\tilde{p}\tilde{q}}{M^2\tilde{p}^2 - \tilde{q}^2}d\tilde{q}$
Stress gradient	$\dfrac{\partial f}{\partial\sigma_{ij}} = \dfrac{2(\lambda - \kappa)pq}{(1 + e_0)(M^2p^2 + q^2)} \times \left[\frac{1}{3}\frac{M^2p^2 - q^2}{2pq}\delta_{ij} + \frac{3(\sigma_{ij} - p\delta_{ij})}{2q}\right]$	$\dfrac{\partial f}{\partial\tilde{\sigma}_{ij}} = \dfrac{2(\lambda - \kappa)\tilde{p}\tilde{q}}{(1 + e_0)(M^2\tilde{p}^2 + \tilde{q}^2)} \times \left[\frac{1}{3}\frac{M^2\tilde{p}^2 - \tilde{q}^2}{2\tilde{p}\tilde{q}}\delta_{ij} + \frac{3(\tilde{\sigma}_{ij} - \tilde{p}\delta_{ij})}{2\tilde{q}}\right]$
Model parameter	$\phi, \lambda/(1 + e_0), \kappa/(1 + e_0), \nu$	

Notes
e_0: initial void ratio; ϕ: angle of internal friction; λ: compression index; κ: swelling index.

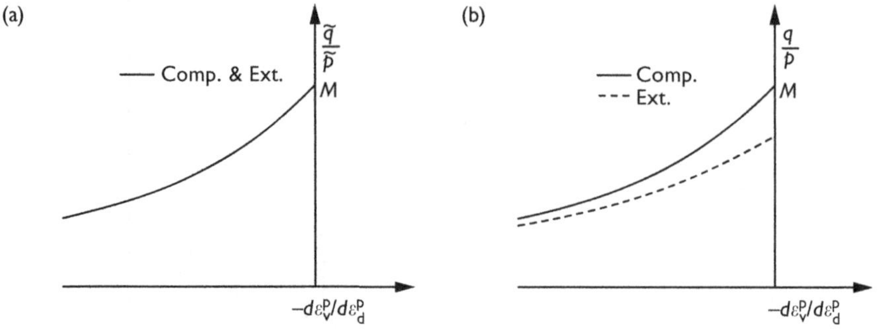

Figure 3.10 Stress–dilatancy relationships of the revised modified Cam-clay model under triaxial compression and triaxial extension.

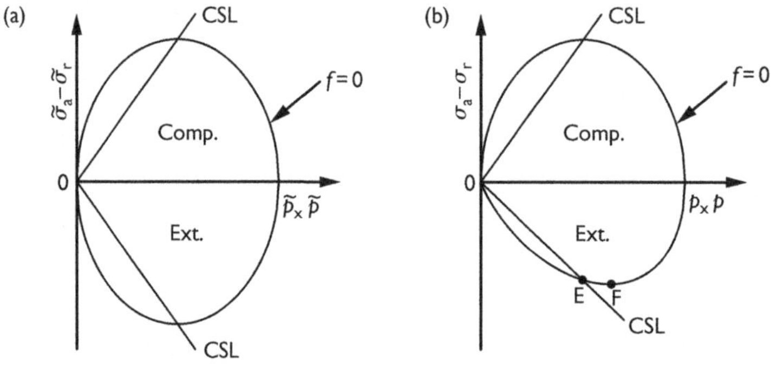

Figure 3.11 Yield loci of modified Cam-clay model generalized by the SMP criterion in triaxial compression and triaxial extension.

3 The soil parameters of the revised models are completely the same as those of the original models because $\tilde{\sigma}_{ij} = \sigma_{ij}$ in triaxial compression stress. On the other hand, if a model is revised by t_{ij}, the soil parameters need to be changed.
4 The stress condition at the critical state in the revised models, which is expressed by the Eq. (3.13), satisfies the SMP failure criterion. On the other hand, the stress condition at the critical state in the t_{ij}-clay model does not satisfy the SMP failure criterion.

3.4 Comparison of model predictions with experimental data

The capability of the four models (the original and modified Cam-clay models and the corresponding models revised by the SMP criterion) in predicting drained and undrained behavior of clay is examined here. The experimental data on the normally consolidated Fujinomori clay under triaxial compression, triaxial extension, plane strain, and true triaxial conditions (Nakai and Matsuoka 1986; Nakai *et al.* 1986) are used for the examination. The values of the soil parameters used in the model predictions for Fujinomori clay are listed in

Table 3.3 Model parameters

$\lambda/(1+e_0)$	$\kappa/(1+e_0)$	ϕ'	ν
0.0508	0.0112	33.7	0.3

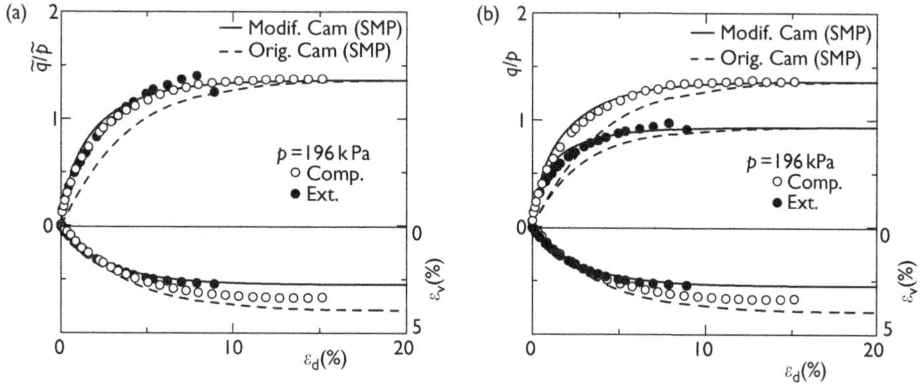

Figure 3.12 Predicted and measured results of triaxial compression and triaxial extension tests on Fujinomori clay.

Table 3.3. These values are the same as the ones the original authors used, except for Poisson's ratio. Poisson's ratio is assumed to be 0.3 in this chapter. In Figures 3.12–3.16, the sign ○ represents the measured results, the abbreviations Orig. Cam and Modif. Cam represent the results predicted by the original and modified Cam-clay models, respectively, and the abbreviations Orig. Cam (SMP) and Modif. Cam (SMP) represent the results predicted by the original and modified Cam-clay models revised by SMP criterion, respectively.

Figure 3.12 shows the predicted and experimental results of Fujinomori clay in triaxial compression and triaxial extension under $p = 196$ kPa, plotted in terms of (a) $\tilde{q}/\tilde{p} \sim \varepsilon_d \sim \varepsilon_v$ and (b) $q/p \sim \varepsilon_d \sim \varepsilon_v$. The curves in Figure 3.12 are the results predicted by the revised Cam-clay models. The data obtained from triaxial compression and triaxial extension tests are represented by one single curve in the $\tilde{q}/\tilde{p} \sim \varepsilon_d$ plane, but by two separate curves in the $q/p \sim \varepsilon_d$ plane. It can be seen that the revised models predict these characteristics of the stress–strain relations.

Figure 3.13 shows the predicted and experimental results for the drained behavior of Fujinomori clay in triaxial compression and triaxial extension at $\sigma_3 = 196$ kPa, plotted in terms of the major and minor principal stress ratio σ_1/σ_3, the major principal strain ε_1 and the volumetric strain ε_v. It can be seen from Figure 3.13(a) that the results predicted by the Cam-clay models and the revised counterparts are the same in triaxial compression. This is because the transformed stress $\tilde{\sigma}_{ij}$ is equal to the conventional stress σ_{ij} in triaxial compression. It can be seen from Figure 3.13(b) that the results predicted by the revised Cam-clay models fit the experimental data better than those predicted by their counterparts. This is because the revised models can account for the differences of the shear strength and the shear yield of clay between triaxial compression and triaxial extension.

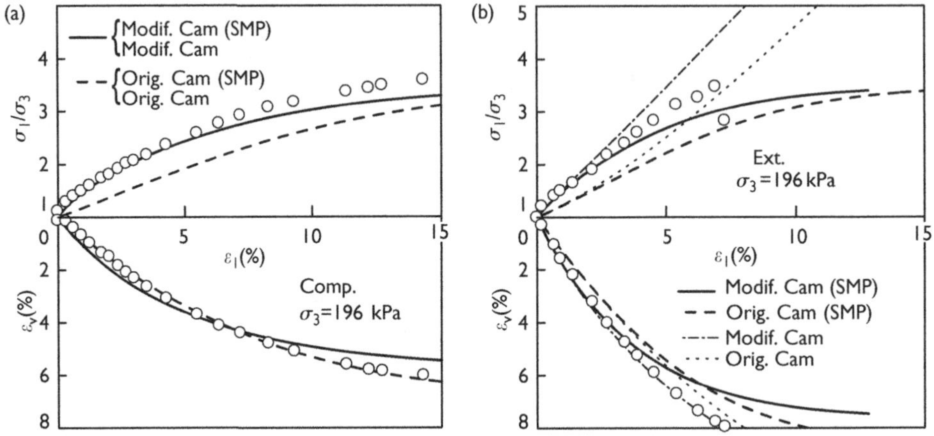

Figure 3.13 Comparison between predicted and measured stress–strain behavior ($\sigma_3 = 196$ kPa).

Figure 3.14 shows the predicted results and the results from true triaxial test ($\theta = 0°$, $\theta = 15°$, $\theta = 30°$, $\theta = 45°$, and $\theta = 60°$) on the drained behavior of Fujinomori clay at $p = 196$ kPa, plotted in terms of the principal stress ratio σ_1/σ_3, the principal strains ($\varepsilon_1, \varepsilon_2$, and ε_3), and the volumetric strain ε_v. It can be seen from Figure 3.14(a) that the predicted results by the Cam-clay models and the revised counterparts are the same in triaxial compression. The modified Cam-clay model predicts the experimental data better than the original Cam-clay model for Fujinomori clay. As the stress condition changes from triaxial compression ($\theta = 0°$) to general stress conditions ($\theta > 0°$), the Cam-clay models overpredict the shear strength due to the Extended Mises criterion and cannot predict well the experimental stress–strain behavior. However, the revised Cam-clay models can predict well the stress-strain behavior of clay, as shown in Figure 3.14(b)–(e). It can also be seen from Figure 3.14 that, since the results predicted by the modified Cam-clay model agree well with the results from triaxial compression tests, the results predicted by the revised modified Cam-clay model also agree well with the test results under the other stress conditions ($\theta > 0°$, see Fig. 3.14(b)–(e)). Therefore, if an elastoplastic model, which uses p and q or p and $\eta(= q/p)$ as stress variables, can predict results from triaxial compression tests, it is reasonable to assume that the corresponding model revised by the SMP criterion can predict the stress–strain behavior under the general stress conditions when p and q or p and $\eta(= q/p)$ are replaced with $\tilde{p}$ and $\tilde{q}$ or $\tilde{p}$ and $\tilde{\eta}(= \tilde{q}/\tilde{p})$, respectively.

Figures 3.15 and 3.16 show the predicted and test results of undrained triaxial compression and triaxial extension tests on Fujinomori clay. Figure 3.15 shows the stress–strain behavior under undrained triaxial compression and triaxial extension, plotted in terms of the normalized deviator stress q/p_0 (p_0: initial confining pressure equal to 196 kPa) and the deviator strain ε_d. The revised Cam-clay models can predict the difference in the $q/p_0 \sim \varepsilon_d$ relations between triaxial compression and triaxial extension, but the Cam-clay models cannot. Figure 3.16 shows the effective stress paths for undrained triaxial compression and triaxial extension tests, plotted in terms of the normalized stresses q/p_0 and p/p_0. It can be seen from Figure 3.16(b) that the stress path predicted by the Cam-clay models is higher than that predicted by their

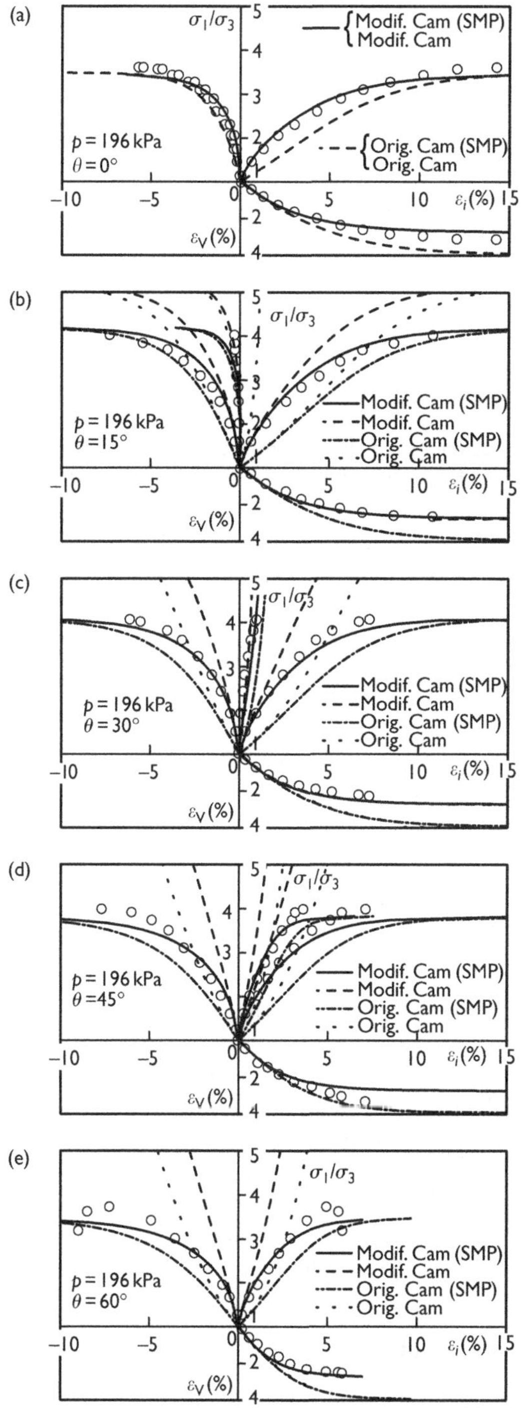

Figure 3.14 Comparison between predicted and measured stress–strain behavior under three different principal stresses (Data after Nakai *et al.* 1986).

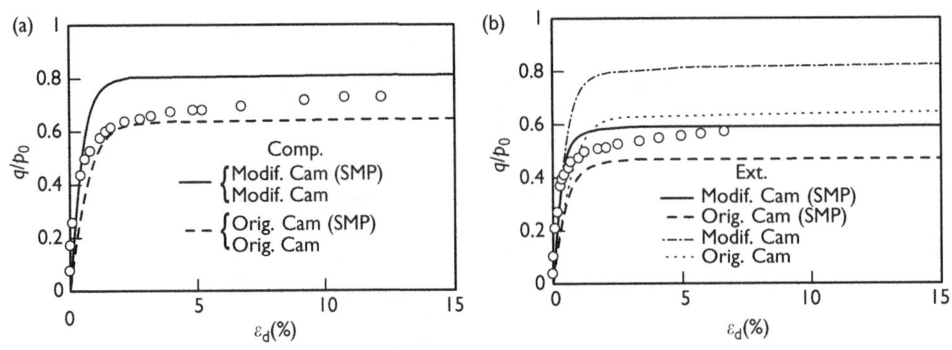

Figure 3.15 Comparison between predicted and measured stress–strain behavior under undrained (a) triaxial compression and (b) triaxial extension (data after Nakai and Matsuoka 1986).

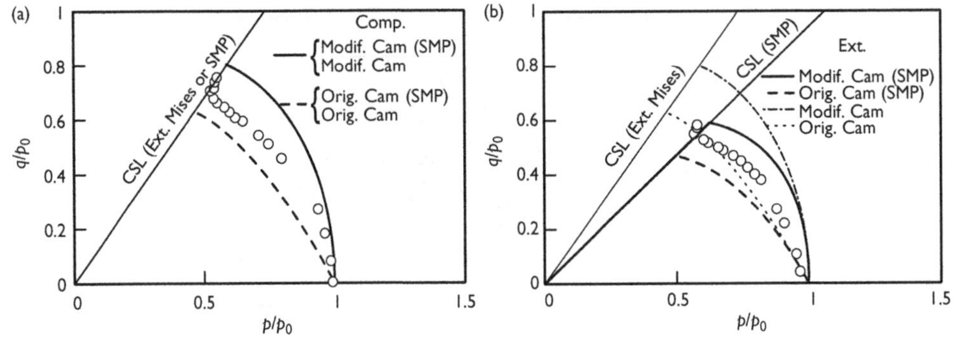

Figure 3.16 Comparison between predicted and measured undrained stress paths under (a) triaxial compression and (b) triaxial extension (data after Nakai and Matsuoka 1986).

revised counterparts under triaxial extension, while the stress paths predicted by the Cam-clay models and their revised counterparts are the same under triaxial compression.

It is clear from Figures 3.12–3.16 that the revised Cam-clay models better predict the experimental data than the Cam-clay models.

From the preceding comparison between the model predictions and the experimental data from element tests on clay, it can be concluded that the revised models are good for predicting deformation and strength of clay in three-dimensional stresses. In order to apply the revised models to practical engineering by the finite element method, we need the elastoplastic constitutive tensor D_{ijkl} and the derivation of the tensor is given in Appendix 2.

3.5 Concluding remarks

The main results are summarized as follows:

1 A transformed stress tensor $\tilde{\sigma}_{ij}$ has been proposed by comparing the shapes of the Extended Mises criterion and the SMP criterion in the principal stress space. The SMP

surface becomes a cone with the axis being the space diagonal in the transformed principal stress space, and a circle with the center being the origin in the transformed π-plane.

2 As an example, the transformed stress tensor $\tilde{\sigma}_{ij}$ is applied to the Cam-clay models to combine the critical state theory with the SMP criterion. The revised Cam-clay models keep the consistency between shear yield and shear failure of soils, which means that both yielding and failure obey the SMP criterion. The revised models, with the same soil parameters as the Cam-clay models, are capable of predicting the drained and undrained behavior of clay in general stress conditions. The comparison between the predictions of the revised Cam-clay model and the experimental results indicates a good performance of the models in predicting the behavior of clay not only under triaxial compression but also under triaxial extension and true triaxial conditions.

3 Throughout this chapter, we find that, since the results predicted by the modified Cam-clay model agree well with the experimental results for Fujinomori clay in triaxial compression, the results predicted by the revised modified Cam-clay model also agree well with the experimental results under other stress conditions. Therefore, it is possible to extend any other elastoplastic models for soils, whose validity has been verified only in triaxial compression condition, to the general stress conditions by simply replacing the ordinary stress tensor σ_{ij} with the transformed stress tensor $\tilde{\sigma}_{ij}$.

Appendix I Derivation of ℓ_0

As shown in Figure 3.1(b) and Figure 3.4(a), when τ_{SMP}/σ_{SMP} and p are given, ℓ_0 can be determined. Under triaxial compression condition ($\sigma_1 > \sigma_2 = \sigma_3$), the following equations can be obtained.

$$\ell_0 = \sqrt{\sigma_1^2 + 2\sigma_3^2 - \left(\frac{\sigma_1 + 2\sigma_3}{\sqrt{3}}\right)^2} = \sqrt{\frac{2}{3}}(\sigma_1 - \sigma_3) \quad \text{(A3.1.1)}$$

In order to express ℓ_0 as a function of the mean stress $p(p = (\sigma_1 + 2\sigma_3)/3)$ and the mobilized internal friction angle ϕ_{mob} or the stress ratio τ_{SMP}/σ_{SMP}, Eq. (A3.1.1) is rewritten as

$$\ell_0 = \sqrt{\frac{2}{3}}\frac{\sigma_1 - \sigma_3}{p}p = \sqrt{\frac{2}{3}}\frac{3(\sigma_1 - \sigma_3)}{\sigma_1 + 2\sigma_3}p \quad \text{(A3.1.2)}$$

As shown in Figure A3.1.1, the relationship between the mobilized internal friction angle ϕ_{mob} and the principal stresses σ_1 and σ_3 under triaxial compression can be given as follows:

$$\sin\phi_{mob} = \frac{\sigma_1 - \sigma_3}{\sigma_1 + \sigma_3} \quad \text{or} \quad \frac{\sigma_1}{\sigma_3} = \frac{1 + \sin\phi_{mob}}{1 - \sin\phi_{mob}} \quad \text{(A3.1.3)}$$

Under triaxial compression and extension conditions, τ_{SMP}/σ_{SMP} can be expressed as follows:

$$\frac{\tau_{SMP}}{\sigma_{SMP}} = \frac{\sqrt{2}}{3}\left(\sqrt{\frac{\sigma_1}{\sigma_3}} - \sqrt{\frac{\sigma_3}{\sigma_1}}\right) = \frac{2\sqrt{2}}{3}\frac{\sigma_1 - \sigma_3}{2\sqrt{\sigma_1\sigma_3}} = \frac{2\sqrt{2}}{3}\tan\phi_{mob} \quad \text{(A3.1.4)}$$

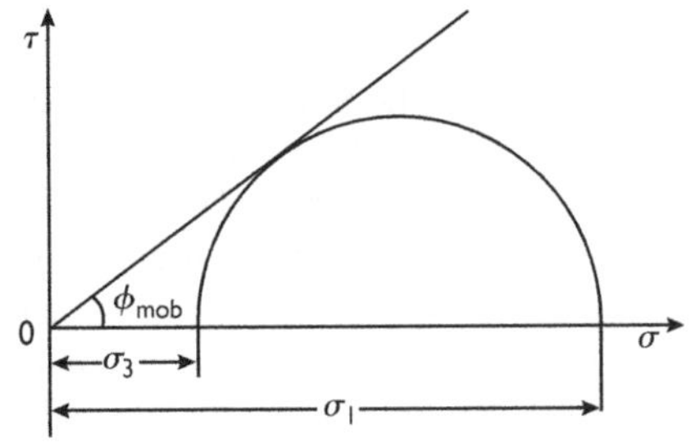

Figure A3.1.1

Combining Eqs (A3.1.2), (A3.1.3), and (A3.1.4) gives

$$\ell_0 = \sqrt{\frac{2}{3}}\frac{6\sin\phi_{\text{mob}}}{3-\sin\phi_{\text{mob}}}p = \sqrt{\frac{2}{3}}\frac{6p}{3\sin^{-1}\phi_{\text{mob}}-1} = \sqrt{\frac{2}{3}}\frac{6p}{3\sqrt{1+\tan^{-2}\phi_{\text{mob}}}-1}$$

$$= \sqrt{\frac{2}{3}}\frac{6p}{3\sqrt{1+8(\tau_{\text{SMP}}/\sigma_{\text{SMP}})^{-2}/9}-1}$$

$$= \sqrt{\frac{2}{3}}\frac{2I_1}{3\sqrt{(I_1I_2-I_3)/(I_1I_2-9I_3)}-1} \qquad \text{(A3.1.5)}$$

Appendix 2 Derivation of elastoplastic constitutive tensor D_{ijkl}

The elastoplastic constitutive tensor of the revised model used in solving boundary value problems by means of the finite element method is derived here. The elastic part of the stress–strain relation can be written in the incremental form as

$$d\sigma_{ij} = D^{\text{e}}_{ijkl}d\varepsilon^{\text{e}}_{kl} = D^{\text{e}}_{ijkl}(d\varepsilon_{kl} - d\varepsilon^{\text{p}}_{kl}) \qquad \text{(A3.2.1)}$$

Where ε_{kl} and $\varepsilon^{\text{p}}_{kl}$ are the total strain tensor and its plastic component, respectively. D^{e}_{ijkl} is the elastic constitutive tensor, and can be found from generalized Hooke's law for isotropic materials,

$$D^{\text{e}}_{ijkl} = L\delta_{ij}\delta_{kl} + G(\delta_{ik}\delta_{jl} + \delta_{il}\delta_{jk}) \qquad \text{(A3.2.2)}$$

where L and G are Lame's constants, which are written from $e = e_0 - \kappa \ln p/p_0$ in the elastic region by

$$G = \frac{E}{2(1+\nu)} = \frac{3(1-2\nu)(1+e_0)}{2(1+\nu)\kappa}p \qquad \text{(A3.2.3)}$$

$$L = \frac{E}{3(1-2\nu)} - \frac{2}{3}G = \frac{(1+e_0)}{\kappa}p - \frac{2}{3}G \qquad \text{(A3.2.4)}$$

where E and ν are the elastic modulus and Poisson's ratio, respectively.

As shown in Tables 3.1 and 3.2, the yield function f is a function of the transformed stress tensor and the hardening parameter (the plastic volumetric strain ε_v^p), i.e.

$$f = f_1(\tilde{\sigma}_{ij}) - \varepsilon_v^p = 0 \tag{A3.2.5}$$

Substituting Eq. (3.12) into Eq. (A3.2.5) gives

$$f = f_2(\sigma_{ij}) - \varepsilon_v^p = 0 \tag{A3.2.6}$$

The consistency condition can be written as

$$df = \frac{\partial f}{\partial \sigma_{ij}} d\sigma_{ij} + \frac{\partial f}{\partial \varepsilon_v^p} d\varepsilon_v^p = 0 \tag{A3.2.7}$$

Substituting Eqs (A3.2.1) and (3.15) into Eq. (A3.2.7) gives

$$\frac{\partial f}{\partial \sigma_{ij}} D_{ijkl}^e \left(d\varepsilon_{kl} - \Lambda \frac{\partial f}{\partial \tilde{\sigma}_{kl}} \right) + \frac{\partial f}{\partial \varepsilon_v^p} \Lambda \frac{\partial f}{\partial \tilde{\sigma}_{ij}} \delta_{ij} = 0 \tag{A3.2.8}$$

From Eq. (A3.2.6) we have $\partial f / \partial \varepsilon_v^p = -1$. Rearranging (A3.2.8) gives

$$\Lambda = \frac{(\partial f / \partial \sigma_{ij}) D_{ijkl}^e d\varepsilon_{kl}}{X} \tag{A3.2.9}$$

where

$$X = \frac{\partial f}{\partial \tilde{\sigma}_{ij}} \delta_{ij} + \frac{\partial f}{\partial \sigma_{ij}} D_{ijkl}^e \frac{\partial f}{\partial \tilde{\sigma}_{kl}} \tag{A3.2.10}$$

Substituting Eqs (3.15) and (A3.2.9) into Eq. (A3.2.1), we can obtain a general form of the proposed model.

$$d\sigma_{ij} = D_{ijkl} d\varepsilon_{kl} \tag{A3.2.11}$$

where the elastoplastic constitutive tensor is

$$D_{ijkl} = D_{ijkl}^e - D_{ijmn}^e \frac{\partial f}{\partial \tilde{\sigma}_{mn}} \frac{\partial f}{\partial \sigma_{st}} D_{stkl}^e / X \tag{A3.2.12}$$

Introducing the elastic constitutive tensor of Hooke's law for isotropic elasticity into Eq. (A3.2.10) and (A3.2.12), we obtain

$$X = \frac{\partial f}{\partial \tilde{\sigma}_{ii}} + L \frac{\partial f}{\partial \sigma_{ii}} \frac{\partial f}{\partial \tilde{\sigma}_{jj}} + 2G \frac{\partial f}{\partial \sigma_{ij}} \frac{\partial f}{\partial \tilde{\sigma}_{ij}} \tag{A3.2.13}$$

$$D_{ijkl} = L\delta_{ij}\delta_{kl} + G(\delta_{ik}\delta_{jl} + \delta_{il}\delta_{jk}) - \left(L \frac{\partial f}{\partial \tilde{\sigma}_{mm}} \delta_{ij} + 2G \frac{\partial f}{\partial \tilde{\sigma}_{ij}} \right) \times \left(L \frac{\partial f}{\partial \sigma_{nn}} \delta_{kl} + 2G \frac{\partial f}{\partial \sigma_{kl}} \right) / X \tag{A3.2.14}$$

Equations (A3.2.12) and (A3.2.14) are the generalized expressions of the elastoplastic constitutive model in the incremental form, and can be easily incorporated into finite element computation. Equation (A3.2.12) can also be expressed in the matrix form by

$$\begin{Bmatrix} d\sigma_{11} \\ d\sigma_{22} \\ d\sigma_{33} \\ d\sigma_{12} \\ d\sigma_{23} \\ d\sigma_{31} \end{Bmatrix} = \begin{bmatrix} D_{1111} & D_{1122} & D_{1133} & D_{1112} & D_{1123} & D_{1131} \\ D_{2211} & D_{2222} & D_{2233} & D_{2212} & D_{2223} & D_{2231} \\ D_{3311} & D_{3322} & D_{3333} & D_{3312} & D_{3323} & D_{3331} \\ D_{1211} & D_{1222} & D_{1233} & D_{1212} & D_{1223} & D_{1231} \\ D_{2311} & D_{2322} & D_{2333} & D_{2312} & D_{2323} & D_{2331} \\ D_{3111} & D_{3122} & D_{3133} & D_{3112} & D_{3123} & D_{3131} \end{bmatrix} \begin{Bmatrix} d\varepsilon_{11} \\ d\varepsilon_{22} \\ d\varepsilon_{33} \\ d\gamma_{12} \\ d\gamma_{23} \\ d\gamma_{31} \end{Bmatrix} \tag{A3.2.15}$$

where $\gamma_{ij}(= 2\varepsilon_{ij}, i \neq j)$ is the engineering shear strain. Under plane strain condition, we have $d\sigma_{23} = d\sigma_{31} = 0$ and $d\varepsilon_{33} = d\gamma_{23} = d\gamma_{31} = 0$. When the plane strain condition is applied to Eq. (A3.2.15), we have

$$\begin{Bmatrix} d\sigma_{11} \\ d\sigma_{22} \\ d\sigma_{33} \\ d\sigma_{12} \end{Bmatrix} = \begin{bmatrix} D_{1111} & D_{1122} & D_{1112} \\ D_{2211} & D_{2222} & D_{2212} \\ D_{3311} & D_{3322} & D_{3312} \\ D_{1211} & D_{1222} & D_{1212} \end{bmatrix} \begin{Bmatrix} d\varepsilon_{11} \\ d\varepsilon_{22} \\ d\gamma_{12} \end{Bmatrix} \tag{A3.2.16}$$

In implementing Eqs (A3.2.12) and (A3.2.14) into the finite element program, it is necessary to calculate $\partial f/\partial\tilde{\sigma}_{ij}$ and $\partial f/\partial\sigma_{ij}$. About $\partial f/\partial\tilde{\sigma}_{ij}$, see Tables 3.1 and 3.2. The calculation of $\partial f/\partial\sigma_{ij}$ is as follows:

From the chain rule of differentiation, we have

$$\frac{\partial f}{\partial \sigma_{ij}} = \frac{\partial f}{\partial \tilde{\sigma}_{kl}} \frac{\partial \tilde{\sigma}_{kl}}{\partial \sigma_{ij}} \tag{A3.2.17}$$

From Eq. (3.12),

$$\frac{\partial \tilde{\sigma}_{kl}}{\partial \sigma_{ij}} = \frac{\partial (p\delta_{kl})}{\partial \sigma_{ij}} + \frac{\partial}{\partial \sigma_{ij}} \left(\frac{s_{kl}}{\ell_\theta} \right) \ell_0 + \frac{s_{kl}}{\ell_\theta} \frac{\partial \ell_0}{\partial \sigma_{ij}} \tag{A3.2.18}$$

By considering Eq. (3.3) and $\ell_\theta = \sqrt{s_{ij}s_{ij}}$, Eq. (A3.2.18) can be written as

$$\frac{\partial \tilde{\sigma}_{kl}}{\partial \sigma_{ij}} = \frac{1}{3}\delta_{ij}\delta_{kl} + \left(\delta_{ik}\delta_{jl} - \frac{1}{3}\delta_{ij}\delta_{kl} - \frac{s_{kl}s_{ij}}{\ell_\theta^2} \right) \frac{\ell_0}{\ell_\theta} + \frac{s_{kl}}{\ell_\theta} \frac{\partial \ell_0}{\partial I_m} \frac{\partial I_m}{\partial \sigma_{ij}} \tag{A3.2.19}$$

where I_m ($m = 1, 2$ and 3) is the stress invariant. Therefore, $\partial \ell_0/\partial I_m$ can be easily calculated using Eq. (3.3).

References

Lade, P.V. and Duncan, J.M. 1975, Elastoplastic stress-strain theory for cohesionless soil. *Journal of Geotechnical Engineering ASCE*, 101(10): 1037–1053.

Matsuoka, H. 1976. On the significance of the spatial mobilized plane. *Soil and Foundations*, 16(1): 91–100.

Matsuoka, H. and Nakai, T. 1974. Stress-deformation and strength characteristics of soil under three difference principal stresses. *Proceedings of the JSCE*, 232: 59–70.

Matsuoka, H., Yao, Y.P. and Sun, D.A. 1999. The Cam-clay model revised by the SMP criterion. *Soils and Foundations*, 39(1): 81–95.

Mitachi, T. and Kitago, S. 1979. The influence of stress-history and stress system on the stress-strain-strength properties of saturated clay. *Soils and Foundations*, 19(2): 45–61.

Miura, N., Murata, H., and Yasufuku, N. 1984. Stress strain characteristics of sand in a particle-crushing region. *Soils and Foundations*, 24(1): 77–89.

Nakai, T. and Matsuoka, H. 1986. A generalized elastoplastic constitutive model for clay in three-dimensional stresses. *Soils and Foundations*, 26(3): 81–98.

Nakai, T. and Mihara, Y. 1984. A new mechanical quantity for soils and its application to elastoplastic constitutive models. *Soils and Foundations*, 24(2): 82–94.

Nakai, T., Matsuoka, H., Okuno, N., and Tsuzuki, K. 1986. True triaxial tests on normally consolidated clay and analysis of the observed shear behavior using elastoplastic constitutive models. *Soils and Foundations*, 26(4): 67–78.

Nayak, G.C. and Zienkiewicz, O.C. 1972. Elasto-plastic stress analysis – A generalisation for various constitutive relations including strain softening. *International Journal for Numerical Methods in Engineering*, 5: 113–135.

Randolph, M.F. 1982. Generalising the Cam-clay models. Symposium on the Implementation of Critical State Soil Mechanics in F. E. Computations, Cambridge University.

Roscoe, K.H. and Burland, J.B. 1968. *On the Generalised Stress-strain Behavior of "Wet" Clay*. Engineering Plasticity, Cambridge University Press, Cambridge, UK, pp. 535–609.

Roscoe, K.H., Schofield, A.N., and Thurairajah, A. 1963. Yielding of clay in states wetter than critical. *Geotechnique*, 13(3): 211–240.

Schofield, A.N. and Wroth, C.P. 1968. *Critical State Soil Mechanics*. Mcgraw-Hill, London.

Sheng, D., Sloan, S.W., and Yu, H.S. 2000. Aspects of finite element implementation of critical state models. *Computational Mechanics*, 26(2): 185–196.

Van Eekelen, H.A.M. 1980. Isotropic yield surfaces in three dimensions for use in soil mechanics. *International Journal for Numerical and Analytical Methods in Geomechanics*, 8: 259–286.

Wroth, C.P. and Houlsby, G.T. 1985. Soil mechanics-property characterization and analysis procedures. Proceedings of the 11th International Conference on Soil Mechanics and Foundation Engineering, Vol. 1, Francisco, pp. 1–55.

Zienkiewicz, O.C. and Pande, G.N. 1977. Some useful forms of isotropic yield surfaces for soil and rock mechanics. In: G. Gudehus (Ed.): *Finite Elements in Geomechanics*, Chapter 5, London, pp. 179–198.

Chapter 4

Elastoplastic constitutive models for geomaterials using transformed stress

In Chapter 3, we have introduced the transformed stress based on the SMP criterion to generalize the Cam-clay models. The generalized models are essentially applicable to the normally consolidated clay under general stress states, and cannot accurately predict the mechanical behavior of other geomaterials such as sands and unsaturated soils. In this chapter, we further introduce three elastoplastic constitutive models for sands, K_0-consolidated soils, and unsaturated soils.

4.1 An elastoplastic model for sands and clays

In this section, a unified constitutive model for both clay and sand under three-dimensional stress conditions is developed based on the modified Cam-clay model, taking into consideration the following two points (Yao *et al.* 1999). First, a transformed stress tensor based on the SMP criterion is applied to the modified Cam-clay model. The model keeps the consistency between shear yielding and shear failure and is a combination of the critical state theory and the SMP criterion theory. Second, a new hardening parameter is derived in order to develop a unified constitutive model for both clay and sand. This new hardening law can describe not only the dilatancy of sand, but also the volumetric strain hardening of clay. The validity of the hardening parameter is confirmed by experimental results from triaxial compression and extension tests on sand along the various stress paths. Only five conventional soil parameters are needed in the present model.

4.1.1 Introduction

The modified Cam-clay model, which is proposed by Roscoe and Burland (1968) and is suitable for normally consolidated clay, is shortened to the Cam-clay model for simplicity in this section. The term clay is to be interpreted as the normally consolidated clay in this section. The Cam-clay model, as many other models, were initially developed for axisymmetric stress states and have then been generalized to three-dimensional stress states by assuming a circular shape of the yield surface in the deviatoric plane. The mean stress $p(=\sigma_{ii}/3)$ and deviator stress $q(=\sqrt{3s_{ij}s_{ij}/2})$ are used as the stress variables in the model, with $s_{ij}(=\sigma_{ij}-p\delta_{ij})$ being a deviatoric stress tensor and δ_{ij} being the Kronecker delta. That is to say, the Extended Mises criterion ($q/p=$ constant) is adopted for shear yielding and shear failure of clay in the Cam-clay model. However, as the experimental evidence shows in Chapters 1 and 3, the Extended Mises criterion grossly overestimates the shear strength under triaxial extension stress state, and also results in incorrect intermediate stress ratio (σ_2/σ_3) in plane strain condition (Wroth and Houlsby 1985). On the other hand, the SMP failure criterion (Matsuoka and Nakai 1974), which is considered to be a three-dimensional extension of the Mohr-Coulomb criterion, is one of the failure criteria that describe well the recent test results of soils. A transformed stress tensor $\tilde{\sigma}_{ij}$ has been proposed in Chapter 3, which is deduced from the transformation of the SMP criterion to a circle in the transformed deviatoric plane. The generalized models using the transformed stress $\tilde{\sigma}_{ij}$, has realized the consistency from shear yielding to shear failure of soils in three-dimensional stresses, i.e., both of which obey the SMP criterion, and the combination of the critical state theory and the SMP criterion theory.

On the other hand, the plastic volumetric strain $\varepsilon_{\mathrm{v}}^{\mathrm{p}}$ is taken as a hardening parameter in the Cam-clay model, which is not appropriate for dilatant sand. So far, a variety of the hardening laws have been proposed to describe the dilatancy of soils (e.g. Lade 1977; Nova and Wood 1979; Nakai 1989). Several elastoplastic models had been proposed for dilatant sand (e.g. Hashiguchi and Ueno 1977; Pastor *et al.* 1985; Prevost 1985; Zienkiewicz *et al.* 1985; Dafalias 1986). In this section, a new hardening parameter is derived on the basis of the consideration that unified yield and plastic potential functions can be used for both clay and sand. These yield and plastic potential functions are identical to those in the Cam clay model. The proposed hardening parameter can describe not only the dilatancy of sand (from lightly to heavily dilatant sand), but also recover the plastic volumetric strain $\varepsilon_{\mathrm{v}}^{\mathrm{p}}$ for clay. The validity of the hardening parameter $\tilde{H}$ is confirmed by experimental results from triaxial compression and extension tests on sand along the various stress paths.

Finally, the capability of the present model in predicting drained behavior of sand is examined along various stress paths under triaxial compression and extension conditions. The results predicted by the present model agree well with the test results under triaxial compression and triaxial extension stress states. Only five soil parameters are needed in the present model. These parameters can be determined by a loading and unloading isotropic consolidation test and a conventional triaxial compression test.

4.1.2 A unified hardening parameter for both clays and sands

The modified Cam-clay model is considered to be one of the fundamental elastoplastic models for soils, and is considered to be better than the original Cam-clay model (Roscoe *et al.* 1963) for clay. In the modified Cam-clay model, the yield and plastic potential functions are assumed to be in the same form as follows:

$$f = g = c_{\mathrm{p}}\left[\ln\frac{p}{p_0} + \ln\left(1 + \frac{q^2}{M^2p^2}\right)\right] - \varepsilon_{\mathrm{v}}^{\mathrm{p}} = 0 \tag{4.1.1}$$

where p_0 is the initial mean stress, M is the value of q/p at the critical state, $d\varepsilon_{\mathrm{v}}^{\mathrm{p}}(= d\varepsilon_{ii}^{\mathrm{p}})$ is the plastic volumetric strain increment and

$$c_{\mathrm{p}} = \frac{\lambda - \kappa}{1 + e_0} \tag{4.1.2}$$

In the equation here, λ is the compression index, κ is the swelling index and e_0 is the initial void ratio at $p = p_0$. Equation (4.1.1) is also a reasonable choice of the plastic potential function for dilatant sand, because this equation can explain that the plastic volumetric strain increment is negative before the phase transformation ($\eta = M$) (Ishihara *et al.* 1975) and positive after the phase transformation (see Fig. 4.1.1). In order to adopt an associated flow

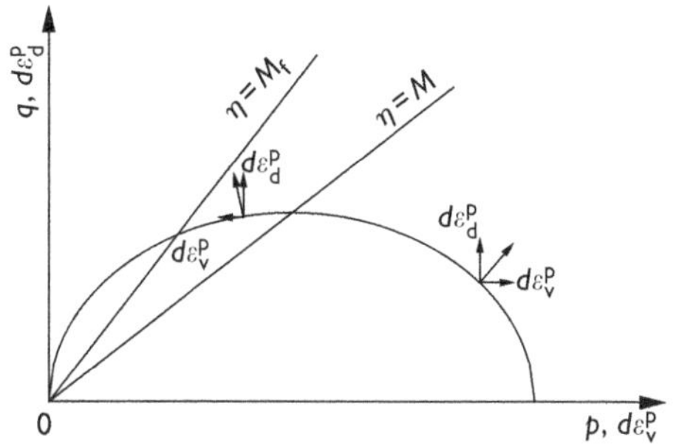

Figure 4.1.1 Direction of plastic strain increment.

rule, the yield function identical to Eq. (4.1.1) is also used for sand. However, the plastic volumetric strain cannot be used as the hardening parameter because it does not increase monotonously with loading (see the test results of the next section for details). In this part, a new hardening parameter H is derived to describe the hardening behavior of clay and sand. The yield and plastic potential functions for sand are written as

$$f = g = c_p \left[\ln \frac{p}{p_0} + \ln \left(1 + \frac{q^2}{M^2 p^2} \right) \right] - H = 0 \tag{4.1.3}$$

The hardening parameter is usually considered to be a combination of the stress tensor σ_{ij} and plastic strain increment tensor $d\varepsilon_{ij}^p$, e.g., the plastic-work-type hardening parameter. Therefore, without losing its generality, we assume that the hardening parameter takes the form of:

$$H = \int dH = \int \left[c_1(\sigma_{ij}) d\varepsilon_v^p + c_2(\sigma_{ij}) d\varepsilon_d^p \right] \tag{4.1.4}$$

where $c_1(\sigma_{ij})$ and $c_2(\sigma_{ij})$ are functions of the stress tensor, $d\varepsilon_d^p$ $(= \sqrt{2(d\varepsilon_{ij}^p - d\varepsilon_v^p \delta_{ij}/3)(d\varepsilon_{ij}^p - d\varepsilon_v^p \delta_{ij}/3)/3})$ is the plastic deviator strain increment. In the modified Cam-clay model, the following stress–dilatancy equation is adopted:

$$\frac{d\varepsilon_v^p}{d\varepsilon_d^p} = \frac{M^2 - \eta^2}{2\eta} \tag{4.1.5}$$

where η is the stress ratio $(= q/p)$.

Substituting Eq. (4.1.5) into Eq. (4.1.4) gives

$$H = \int \left[c_1(\sigma_{ij}) d\varepsilon_v^p + c_2(\sigma_{ij}) \frac{2\eta}{M^2 - \eta^2} d\varepsilon_v^p \right] = \int c(\sigma_{ij}) d\varepsilon_v^p \tag{4.1.6}$$

where $c(\sigma_{ij})$ is a function of the stress tensor. By substituting Eq. (4.1.6) to Eq.(4.1.3), the total differential form of the yield function is expressed as

$$\begin{aligned} df &= \frac{\partial f}{\partial p}dp + \frac{\partial f}{\partial q}dq + \frac{\partial f}{\partial H}dH \\ &= \frac{\partial f}{\partial p}dp + \frac{\partial f}{\partial q}dq - c(\sigma_{ij})d\varepsilon_{\mathrm{v}}^{\mathrm{p}} \\ &= \frac{\partial f}{\partial p}dp + \frac{\partial f}{\partial q}dq - c(\sigma_{ij})\Lambda\frac{\partial f}{\partial p} = 0 \end{aligned} \tag{4.1.7}$$

So the proportionality constant Λ can be written as

$$\Lambda = \frac{1}{c(\sigma_{ij})}\frac{(\partial f/\partial p)dp + (\partial f/\partial q)dq}{(\partial f/\partial p)} \tag{4.1.8}$$

From Eq. (4.1.3), the following two differential equations can be obtained:

$$\frac{\partial f}{\partial p} = \frac{c_{\mathrm{p}}}{p}\frac{M^2 - \eta^2}{M^2 + \eta^2} \tag{4.1.9}$$

$$\frac{\partial f}{\partial q} = \frac{c_{\mathrm{p}}}{p}\frac{2\eta}{M^2 + \eta^2} \tag{4.1.10}$$

By substituting Eqs (4.1.9) and (4.1.10) into Eq. (4.1.8), the plastic deviator strain increment along the constant mean stress path ($dp = 0$) can be expressed as follows

$$d\varepsilon_{\mathrm{d}}^{\mathrm{p}} = \Lambda\frac{\partial f}{\partial q} = \frac{1}{c(\sigma_{ij})}\frac{c_{\mathrm{p}}}{p}\frac{4\eta^2}{M^4 - \eta^4}dq \tag{4.1.11}$$

Figure 4.1.2 shows the results of triaxial compression tests on Fujinomori clay and Toyoura sand, arranged in (a) $\eta \sim \varepsilon_{\mathrm{d}}$ and (b) $(M_{\mathrm{f}}^4 - \eta^4)/(4\eta^2) \sim d\eta/d\varepsilon_{\mathrm{d}}$ (data from Nakai and Matsuoka 1986). It can be seen from Figure 4.1.2(a) that the shapes of the curves $\eta \sim \varepsilon_{\mathrm{d}}$ for clay and sand are alike. The stress ratios (q/p) at failure are M and M_{f} for clay and sand respectively. In fact, M_{f} is not constant during shearing. Here, M_{f} is simply assumed to be a constant to focus on the dilatancy behavior of sand. Therefore, compared to Eq. (4.1.12) which defines the plastic deviator strain increment for clay in the Cam-clay model under constant mean stress, the plastic deviator strain increment for sand under constant mean stress is assumed to take form as Eq. (4.1.13)

$$d\varepsilon_{\mathrm{d}}^{\mathrm{p}} = c_{\mathrm{p}}\frac{1}{p}\frac{4\eta^2}{M^4 - \eta^4}dq \quad \text{(for clay)} \tag{4.1.12}$$

$$d\varepsilon_{\mathrm{d}}^{\mathrm{p}} = \rho\frac{1}{p}\frac{4\eta^2}{M_{\mathrm{f}}^4 - \eta^4}dq \quad \text{(for sand)} \tag{4.1.13}$$

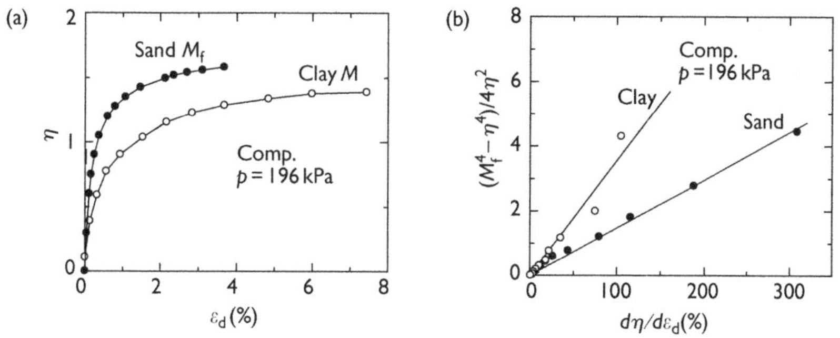

Figure 4.1.2 Triaxial compression test results for clay and sand.

where ρ is a constant. Rearranging (4.1.12) and (4.1.13) gives the following linear forms respectively.

$$\frac{M^4 - \eta^4}{4\eta^2} = c_p \frac{d\eta}{d\varepsilon_d^p} \quad \text{(for clay)} \tag{4.1.14}$$

$$\frac{M_f^4 - \eta^4}{4\eta^2} = \rho \frac{d\eta}{d\varepsilon_d^p} \quad \text{(for sand)} \tag{4.1.15}$$

The validity of Eqs (4.1.14) and (4.1.15) is confirmed by the results of triaxial compression tests from Figure 4.1.2(b). So, Eq. (4.1.13) is a rational assumption for the plastic deviator strain increment. It is worth noting that $M_f = M$ for clay and the elastic deviator strain is very small under a constant mean stress state in Figure 4.1.2(b). Combining Eqs (4.1.11) and (4.1.13) gives

$$c(\sigma_{ij}) = \frac{c_p}{\rho} \frac{M_f^4 - \eta^4}{M^4 - \eta^4} \tag{4.1.16}$$

Substituting Eq. (4.1.16) into Eq. (4.1.6) gives

$$H = \int dH = \int \frac{c_p}{\rho} \frac{M_f^4 - \eta^4}{M^4 - \eta^4} d\varepsilon_v^p \tag{4.1.17}$$

When $\eta = 0$ (isotropic compression stress path), Eq. (4.1.17) becomes

$$H = \int dH = \int \frac{c_p}{\rho} \frac{M_f^4}{M^4} d\varepsilon_v^p \tag{4.1.18}$$

and we have $\varepsilon_v^p = c_p \ln(p/p_0)$. In addition, Eq. (4.1.3) becomes $H = c_p \ln(p/p_0)$ when $\eta = 0$. Therefore, the following equation can be obtained.

$$H = \int dH = \int d\varepsilon_v^p \tag{4.1.19}$$

Comparing Eq. (4.1.18) and Eq. (4.1.19), we have

$$\rho = \frac{c_p M_f^4}{M^4} \tag{4.1.20}$$

Finally, we can obtain the following equation of the new hardening parameter for sand by substituting Eq. (4.1.20) into Eq. (4.1.17).

$$H = \int dH = \int \frac{M^4}{M_f^4} \frac{M_f^4 - \eta^4}{M^4 - \eta^4} d\varepsilon_v^p \tag{4.1.21}$$

If $M = M_f$, Eq. (4.1.21) becomes $H = \int dH = \int d\varepsilon_v^p$, which is the same as the hardening parameter for clay in the Cam-clay model. Therefore, we can say that the new hardening parameter (Eq. (4.1.21)) is a unified one for both clay and sand.

In order to make the new hardening parameter applicable not only to triaxial compression stress but also to other general stress states, H is revised by the transformed stress $\tilde{\sigma}_{ij}$ as

$$\tilde{H} = \int d\tilde{H} = \int \frac{M^4}{M_f^4} \frac{M_f^4 - \tilde{\eta}^4}{M^4 - \tilde{\eta}^4} d\varepsilon_v^p \tag{4.1.22}$$

where

$$\tilde{\eta} = \tilde{q}/\tilde{p} \tag{4.1.23}$$

In order to check the validity of $\tilde{H}$ along various stress paths, the values of $\tilde{H}$ were calculated along the stress paths shown in Figure 4.1.3 for triaxial tests on Toyoura sand (data from Nakai, 1989). The values of the mean stress and principal stress ratio are the same ($p = 588$ kPa, $\sigma_1/\sigma_3 = 4$) at points F and F′. The stress path dependency of the proposed hardening parameter $\tilde{H}$ is to be verified through four kinds of triaxial compression tests (paths: ADEF, ABCF, AF, and ABEF) and three kinds of triaxial extension tests (paths: AD′F′, ACF′ and AF′). Figure 4.1.4 shows the variations of $\tilde{H}$ along the seven kinds of stress paths under triaxial compression and extension conditions. It is obvious from Figure 4.1.4 that the values of the hardening parameter $\tilde{H}$ are uniquely determined at the same stress state, regardless of the stress path in triaxial compression and extension and the previous stress history. Therefore, $\tilde{H}$ is a state quantity. Figure 4.1.5 shows the values of $\tilde{H}$ from results of triaxial compression and extension tests on Toyoura sand under constant mean stress p and the contour lines of $\tilde{H}$. It can be seen that the contour lines are close to the yield curves of the Cam-clay model. Therefore, we can employ $\tilde{H}$ as a new hardening parameter for sand with the yield functions similar to that of the Cam-clay model.

How the hardening parameter describes the dilatancy of soil is explained as follows. From Eq. (4.1.21), we can obtain

$$d\varepsilon_v^p = \frac{M_f^4}{M^4} \frac{M^4 - \tilde{\eta}^4}{M_f^4 - \tilde{\eta}^4} d\tilde{H} \tag{4.1.24}$$

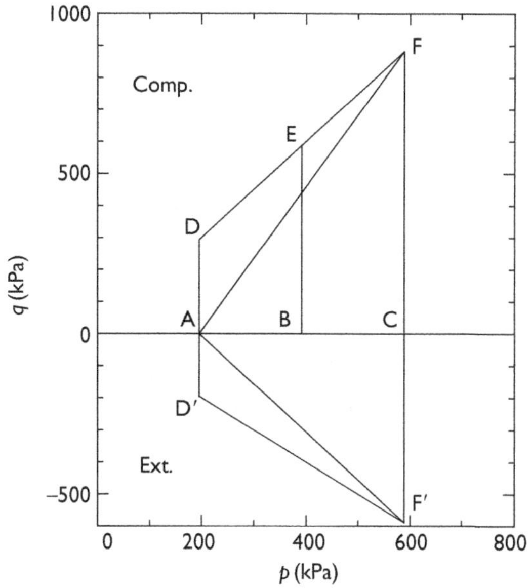

Figure 4.1.3 Stress paths of triaxial tests for examining the new hardening parameter.

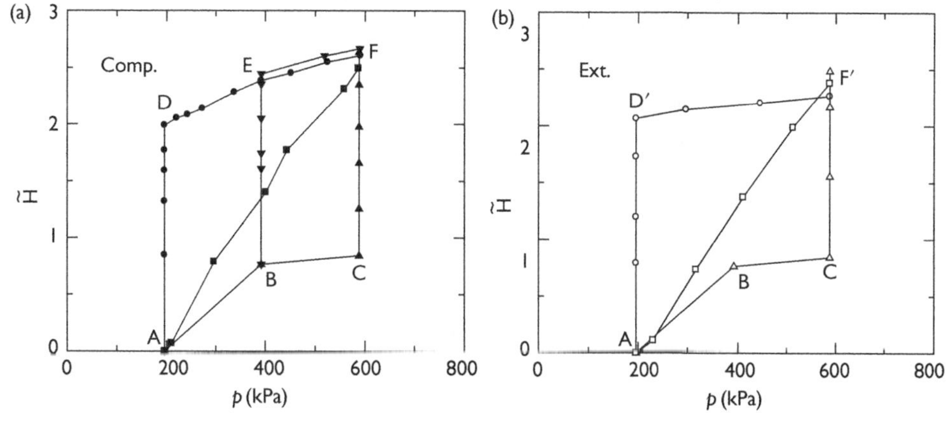

Figure 4.1.4 Relation between the hardening parameter $\tilde{H}$ and mean principal stress p.

Since $\tilde{H}$ is a hardening parameter, $d\tilde{H}$ is always larger than or equal to zero. Taking this into account, the following dilatancy characteristics can be obtained from Eq. (4.1.24):

(1) $0 \leq \tilde{\eta} < M$ (negative dilatancy condition): $d\varepsilon_{\rm v}^{\rm p} > 0$
(2) $\tilde{\eta} = M$ (phase transformation condition): $d\varepsilon_{\rm v}^{\rm p} = 0$
(3) $M < \tilde{\eta} \leq M_{\rm f}$ (positive dilatancy condition): $d\varepsilon_{\rm v}^{\rm p} < 0$

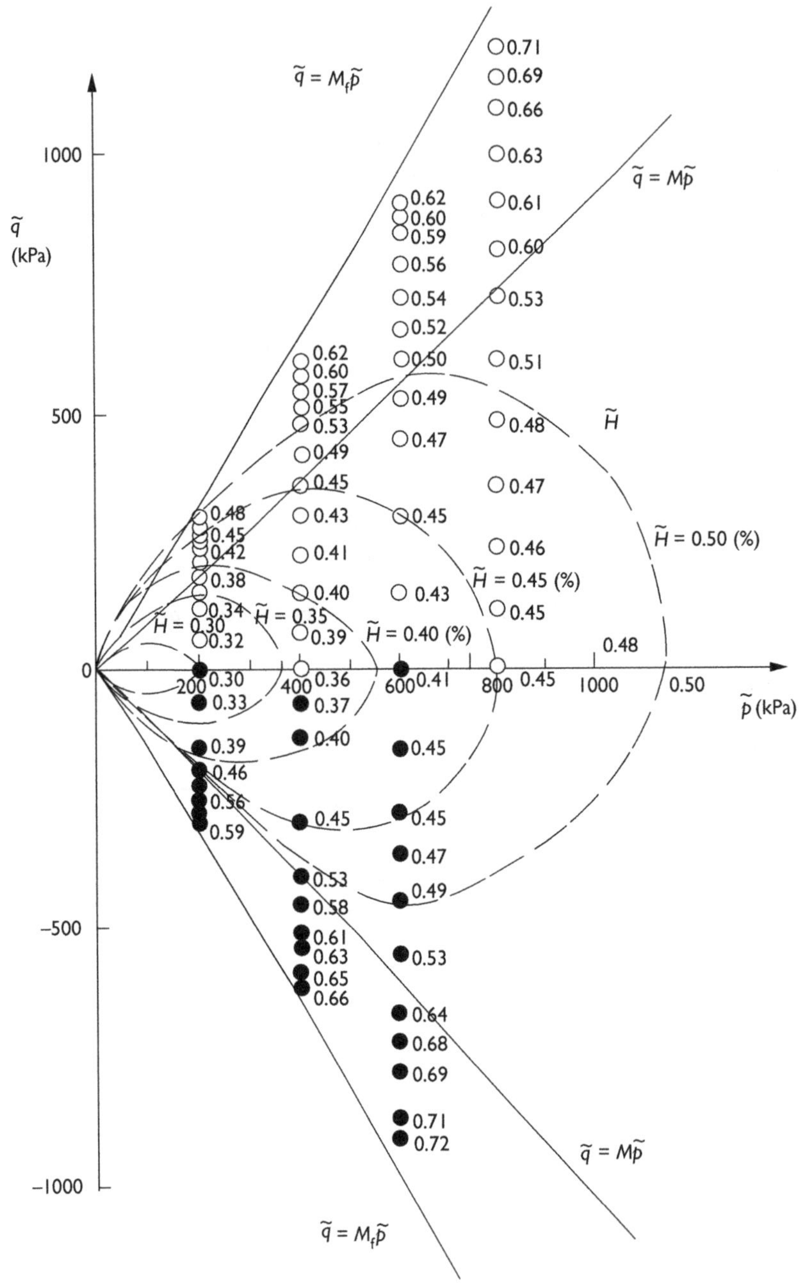

Figure 4.1.5 Values of hardening parameter $\tilde{H}$ along various stress paths and its contour lines.

4.1.3 A unified elastoplastic model for both clays and sands

In the proposed model, the equations of the yield locus and plastic potential remain the same as those of the Cam-clay model, but the transformed stress tensor $\tilde{\sigma}_{ij}$ based on the SMP criterion and the new hardening parameter are adopted to model the mechanical behavior of clay and sand in three-dimensional stresses.

The total strain increment is given by the summation of the elastic and plastic components as usual:

$$d\varepsilon_{ij} = d\varepsilon_{ij}^{\mathrm{e}} + d\varepsilon_{ij}^{\mathrm{p}} \tag{4.1.25}$$

Here, the elastic component is given by Hooke's law:

$$d\varepsilon_{ij}^{\mathrm{e}} = \frac{1+\nu}{E}d\sigma_{ij} - \frac{\nu}{E}d\sigma_{mm}\delta_{ij} \tag{4.1.26}$$

where ν is Poisson's ratio, and the elastic modulus E is expressed as

$$E = \frac{3(1-2\nu)(1+e_0)}{\kappa}p \tag{4.1.27}$$

The plastic component is given by the flow rule in the $\tilde{\sigma}_{ij}$-space:

$$d\varepsilon_{ij}^{\mathrm{p}} = \Lambda\frac{\partial g}{\partial\tilde{\sigma}_{ij}} \tag{4.1.28}$$

where the plastic potential function g (or the yield function f), the hardening parameter $\tilde{H}$, the proportionality constant Λ, and the stress gradient $\partial g/\partial\tilde{\sigma}_{ij}$ are given respectively as follows:

$$f = g = c_{\mathrm{p}}\left[\ln\frac{\tilde{p}}{\tilde{p}_0} + \ln\left(1 + \frac{\tilde{q}^2}{M^2\tilde{p}^2}\right)\right] - \tilde{H} = 0 \tag{4.1.29}$$

$$\tilde{H} = \int d\tilde{H} = \int\frac{M^4}{M_{\mathrm{f}}^4}\frac{M_{\mathrm{f}}^4 - \tilde{\eta}^4}{M^4 - \tilde{\eta}^4}d\varepsilon_{\mathrm{v}}^{\mathrm{p}} \tag{4.1.30}$$

$$\Lambda = \frac{M_{\mathrm{f}}^4}{M^4}\frac{M^4 - \tilde{\eta}^4}{M_{\mathrm{f}}^4 - \tilde{\eta}^4}\left(d\tilde{p} + \frac{2\tilde{p}\tilde{q}}{M^2\tilde{p}^2 - \tilde{q}^2}d\tilde{q}\right) \tag{4.1.31}$$

$$\frac{\partial g}{\partial\tilde{\sigma}_{ij}} = \frac{c_{\mathrm{p}}}{M^2\tilde{p}^2 + \tilde{q}^2}\left[\frac{M^2\tilde{p}^2 - \tilde{q}^2}{3\tilde{p}}\delta_{ij} + 3(\tilde{\sigma}_{ij} - \tilde{p}\delta_{ij})\right] \tag{4.1.32}$$

In these equations, the deviator stress $\tilde{q}$ and the transformed stress ratio $\tilde{\eta}$, M ($\tilde{\eta}$ at the phase transformation) and M_{f} ($\tilde{\eta}$ at peak) are written respectively as follows:

$$\tilde{q} = \sqrt{\frac{3}{2}\tilde{s}_{ij}\tilde{s}_{ij}} \tag{4.1.33}$$

$$M = \frac{6\sin\phi_{\mathrm{pt}}}{3 - \sin\phi_{\mathrm{pt}}} \tag{4.1.34}$$

$$M_{\mathrm{f}} = \frac{6\sin\phi}{3 - \sin\phi} \tag{4.1.35}$$

where ϕ_{pt} is the angle of internal friction at the phase transformation, and ϕ is the angle of internal friction at the shear failure.

4.1.4 Prediction versus experiment

The capability of the proposed model in predicting drained behavior of sand is examined along various stress paths (Fig. 4.1.6) on saturated Toyoura sand (data from Nakai 1989). The material parameters are as follows: $M = 0.95$, $M_f = 1.66$, $\lambda/(1+e_0) = 0.00403$, $\kappa/(1+e_0) = 0.00251$, and $\nu = 0.3$ for Toyoura sand. These soil parameters are determined by the isotropic consolidation test and the conventional triaxial compression test.

Figure 4.1.7 shows the predicted and test results on the drained behavior of Toyoura sand under triaxial compression and extension conditions when $p = 196$ kPa. It can be seen from Figure 4.1.7 that the results predicted by the proposed model (solid lines) agree well with the test results for sand under triaxial compression and extension conditions (marked by ○).

Figure 4.1.8 shows the predicted and test results on the drained behavior of Toyoura sand under the triaxial compression and extension conditions when $\sigma_3 = 196$ kPa. It can be seen from Figure 4.1.8 that the results predicted by the proposed model (solid lines) again agree well with the test results for sand under triaxial compression and extension conditions (marked ○).

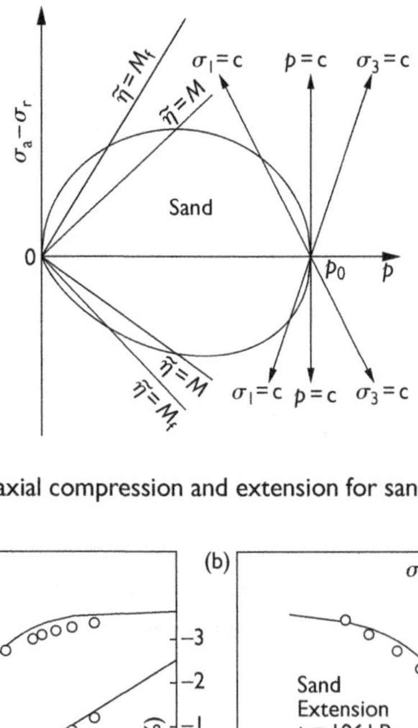

Figure 4.1.6 Stress paths in triaxial compression and extension for sand.

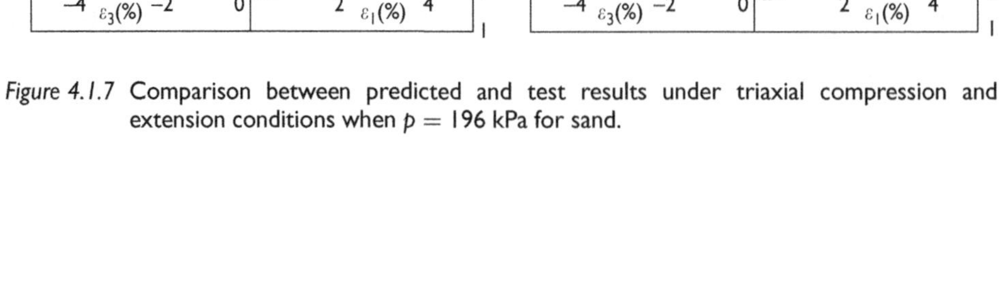

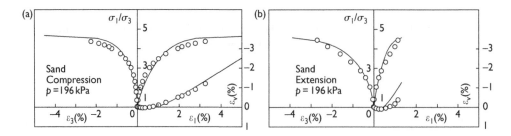

Figure 4.1.7 Comparison between predicted and test results under triaxial compression and extension conditions when $p = 196$ kPa for sand.

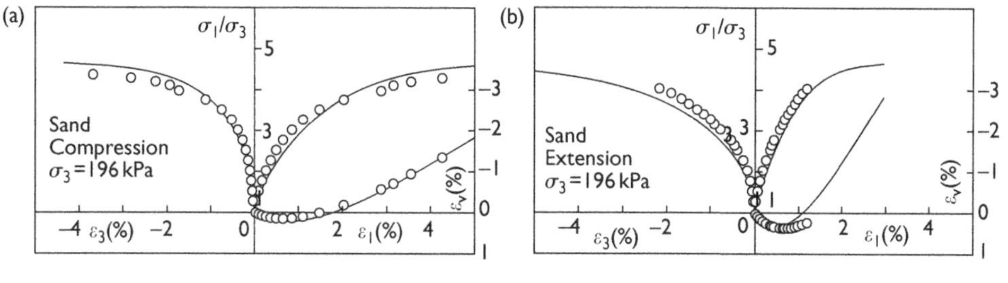

Figure 4.1.8 Comparison between predicted and test results under triaxial compression and extension conditions when $\sigma_3 = 196$ kPa for sand.

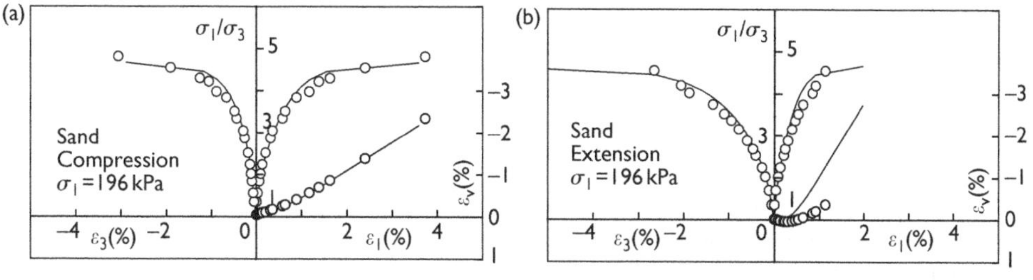

Figure 4.1.9 Comparison between predicted and test results under triaxial compression and extension conditions when $\sigma_1 = 196$ kPa for sand.

Figure 4.1.9 shows the predicted and test results on the drained behavior of Toyoura sand under the triaxial compression and extension conditions when $\sigma_1 = 196$ kPa. The two again agree well with each other.

Therefore, it can be said from the given comparisons in Figures 4.1.7–4.1.9 that the proposed model can reasonably describe the stress–strain response of sand in three-dimensional stresses, and the dilatancy of sand along various stress paths.

The elastoplastic constitutive tensor of the present model required for solving boundary value problems by means of the finite element method can be derived by replacing Equation (A3.2.10) in Chapter 3 with the following equation:

$$X = \frac{M^4(M_f^4 - \tilde{\eta}^4)}{M_f^4(M^4 - \tilde{\eta}^4)}\frac{\partial f}{\partial \tilde{\sigma}_{ii}} + \frac{\partial f}{\partial \sigma_{ij}} D^e_{ijkl} \frac{\partial f}{\partial \tilde{\sigma}_{kl}} \tag{4.1.36}$$

4.1.5 *Modeling confining pressure dependence of strength and deformation*

The elastoplastic model for soil presented in the previous section can describe consistently the deformation and strength characteristics of soils in three-dimensional stresses by introducing a transformed stress $\tilde{\sigma}_{ij}$ into the Cam-clay model, and the dilatancy characteristics of sand by introducing a hardening parameter $\tilde{H}$. However, this model cannot describe the difference of deformation and strength characteristics due to the change of the mean stress. Here a constitutive law for modeling confining-pressure dependency of strength and deformation is proposed (Sun *et al.* 2001). From test results, the confining-pressure dependencies of deformation and strength characteristics of sand can be summarized as follows.

The confining-pressure dependency of strength characteristics means that the peak strength of sand is expressed not by $\tau_f = \sigma \tan\phi$, but by $\tau_f = A\sigma^B$ (A and B are material parameters), i.e., the failure envelope of sand is a gentle convex curve. In some stress range, the failure envelope can be approximately expressed by $\tau_f = c + \sigma \tan\phi$. Therefore, the confining-pressure dependency of strength in three-dimensional stress can be described approximately by the Extended SMP failure criterion, as described in Section 1.5.

The confining-pressure dependency of deformation characteristics means that the stress ratio versus strain curve goes down with increasing confining-pressure, and the positive dilatancy becomes small when the confining pressure increases.

In order to describe this pressure dependency, we need to modify the hardening parameter and the hardening law adopted earlier. The hardening parameter is modified as

$$d\tilde{H} = \frac{M^4}{M_f^4} \cdot \frac{M_f^4 - (\tilde{\tilde{q}}/\tilde{\tilde{p}})^4}{M^4 - (\tilde{\tilde{q}}/\tilde{\tilde{p}})^4} d\varepsilon_v^p \tag{4.1.37}$$

where $\tilde{\tilde{q}}$ and $\tilde{\tilde{p}}$ are the transformed stresses q and p based on the Extended SMP criterion.

In order to match the results of the isotropic compression tests on sand, the volumetric strain under isotropic loading is expressed as

$$\varepsilon_v = C_t \left\{ \left(\frac{p_x}{p_a}\right)^m - \left(\frac{p_0}{p_a}\right)^m \right\} \tag{4.1.38}$$

where C_t and m are material parameters used to indicate the isotropic deformation characteristics, p_0 is an initial mean principal stress, and p_a is the atmospheric pressure.

If the elastic volumetric strain ε_v^e under isotropic loading or unloading is also assumed to be an exponential function of mean principal stress, we have

$$\varepsilon_v^p = \varepsilon_v - \varepsilon_v^e = (C_t - C_e) \left\{ \left(\frac{p_x}{p_a}\right)^m - \left(\frac{p_0}{p_a}\right)^m \right\} \tag{4.1.39}$$

where C_e is a material parameter used to indicate isotropic deformation characteristics. When Eq. (4.1.39) is adopted as the hardening law, the new yield function is written as follows:

$$f = \frac{C_t - C_e}{p_a^m} \left[\left(p + \frac{q^2}{M^2 p}\right)^m - p_0^m \right] - H = 0 \tag{4.1.40}$$

The plots shown in Figure 4.1.10 are the measured results (dots) from triaxial compression and triaxial extension tests on Toyoura sand under $p = 980$ kPa and $p = 1960$ kPa, compared with the predicted results (solid and dotted lines) using the above model that accounts for the confining-pressure dependencies of deformation and strength characteristics. The model prediction shows a good agreement with the measured stress–strain response of the sand.

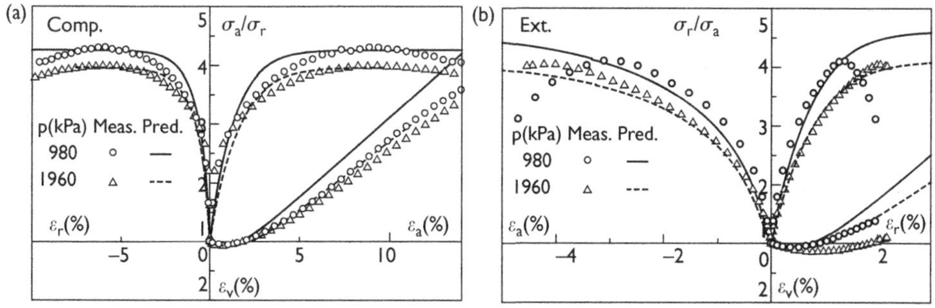

Figure 4.1.10 Predicted and measured results of triaxial compression and extension tests on sand.

4.1.6 Conclusions

1. In this section, a new hardening parameter is derived on the basis of the consideration that unified yield and plastic potential functions can be used for both clay and sand. It can not only describe the dilatancy of lightly to heavily dilatant sand, but also return to the volumetric strain hardening of clay. The physical meaning of this hardening parameter is clear. The validity of the hardening parameter $\tilde{H}$ is confirmed by the experimental results of triaxial compression and extension tests on sand along various stress paths.
2. An elastoplastic model is introduced by applying the transformed stress tensor $\tilde{\sigma}_{ij}$ based on the SMP criterion and the new hardening parameter $\tilde{H}$ to the modified Cam-clay model. The model can reasonably describe the stress–strain–volume behavior of clay and sand in three-dimensional stresses.
3. Five soil parameters (λ, κ, M, M_f, and ν) used in the model can be determined through a loading and unloading isotropic consolidation test and a conventional triaxial compression test.
4. The model can be extended to that predicting well the confining-pressure dependencies of deformation and strength characteristics of sands by some improvements.

4.2 An elastoplastic model for K_0-consolidated soils

4.2.1 Introduction

Sekiguchi and Ohta (1977) proposed an anisotropic hardening elastoplastic model for clays, which can be considered to be an extension of the original Cam-clay model (Roscoe *et al.* 1963) to normally K_0-consolidated clays. This model has been widely applied in the finite element computation to predict the stress and deformation of earth structures in engineering practice in Japan. As the Cam-clay model, the Sekiguchi-Ohta model also uses the Extended Mises criterion as the shear yield and failure criteria, which is known to be inappropriate for soil behavior in three-dimensional (3D) stress states. Instead, criteria such as Matsuoka and Nakai's criterion (the SMP criterion by Matsuoka and Nakai 1974) and Lade's criterion (Lade and Duncan 1975) are known to be more appropriate for shear yielding and failure of soils in 3D stress states. In particular, the SMP criterion is considered to be one of the best criteria for describing shear yield and failure behavior of soils in 3D stress states. In this section, the

Sekiguchi-Ohta model is integrated with the SMP criterion for improving the performance of the model in predicting soil behavior in 3D stress states. Integrating the Sekiguchi-Ohta model with the SMP criterion is achieved by using a transformed stress method, which has been successfully applied to the integration of the Cam-clay models with the SMP criterion as described in Chapter 3. In the revised Sekiguchi-Ohta model, the SMP criterion is taken as the criterion for shear yielding and failure, while the model parameters are the same as those of the original model.

Another shortcoming of the Sekiguchi-Ohta model is that it cannot predict positive dilatancy occurring in medium-to-dense granular materials such as sand. In this section, we first derive the stress–dilatancy relation for the Sekiguchi-Ohta model. The model is then revised by the transformed stress, and by using a new hardening parameter $\tilde{H}$ that can describe the negative and positive dilatancy characteristics of soils with the initially stress-induced anisotropy. All these eventually lead to a unified elastoplastic model for K_0-consolidated clays and sands (Sun *et al.* 2004). The revised model needs only one additional material parameter for practical application, compared to the Sekiguchi-Ohta model.

The aim of this section is to develop a simple anisotropic-hardening elastoplastic model for K_0-consolidated soils including clays, silts, and sands, in which the model parameters are as few as possible and can be determined as easily as possible for practical application. To do that, we use the Sekiguchi-Ohta model as the base and incorporate the SMP criterion and the generalized hardening law described in Chapter 3. The model developed in this section is then compared with experimental data obtained from anisotropically consolidated-drained triaxial shear tests and plane strain tests along different stress paths. The comparison shows that the proposed model gives excellent predictions for normally K_0-consolidated clays and sands. In order to apply the presented simple model to practical engineering, its elastoplastic constitutive tensor is also derived for finite element implementation.

4.2.2 The Sekiguchi-Ohta model and its stress–dilatancy relation

The Sekiguchi-Ohta model is first briefly reviewed for better understanding of the proposed model. The difference between the Sekiguchi-Ohta model and the original Cam-clay model is only the stress ratio used in the models. The Sekiguchi-Ohta model uses the relative stress ratio η^*, which takes into account the initial anisotropic stress state. The yield function (f) and the plastic potential function (g) for the Sekiguchi-Ohta model are expressed as follows:

$$f = g = \frac{\lambda - \kappa}{1 + e_0}\left(\ln\frac{p}{p_0} + \frac{\eta^*}{M}\right) - \varepsilon_v^p = 0 \tag{4.2.1}$$

where λ and κ are the slopes of normal consolidation and swelling lines, M is the slope of the critical state line, e_0 is the void ratio of soil for $p = p_0$, ε_v^p is the plastic volumetric strain that is used as a hardening parameter, and the relative stress ratio η^* is defined as

$$\eta^* = \sqrt{\frac{3}{2}(\eta_{ij} - \eta_{ij0})(\eta_{ij} - \eta_{ij0})} \tag{4.2.2}$$

with

$$\eta_{ij} = \frac{\sigma_{ij} - p\delta_{ij}}{p}, \qquad \eta_{ij0} = \frac{\sigma_{ij0} - p_0\delta_{ij}}{p_0} \tag{4.2.3}$$

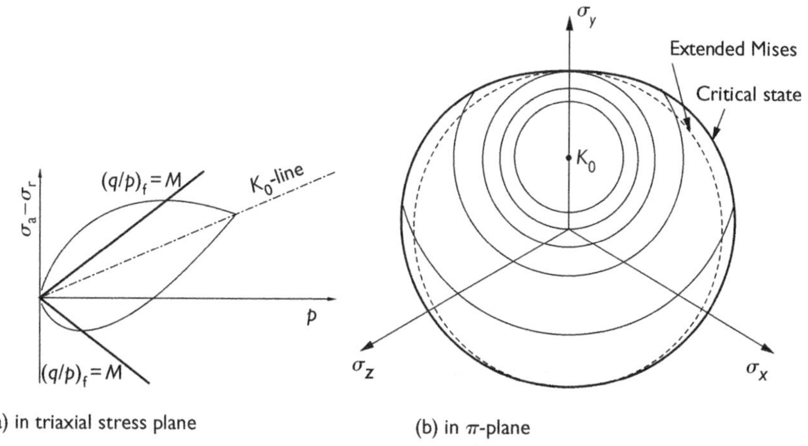

Figure 4.2.1 Yield surface and critical state surface of Sekiguchi-Ohta model.

in which, σ_{ij0} is the value of σ_{ij} at the end of the anisotropic consolidation and δ_{ij} is Kronecker's delta. When shearing starts from the isotropic stress state, we have $\eta^* = \sqrt{3\eta_{ij}\eta_{ij}/2} = q/p = \eta$ because $\eta_{ij0} = 0$, i.e., the Sekiguchi-Ohta model becomes the original Cam-clay model. It is worth noticing that the material parameters (λ, κ, M, e_0, and ν (Poisson's ratio)) in the Sekiguchi-Ohta model are the same as those in the Cam-clay model.

Figure 4.2.1 shows the yield curves (fine solid lines), the critical state lines (bold solid lines) of the Sekiguchi-Ohta model, and the Mises failure criterion ($(q/p)_f = M$) in the $(\sigma_a - \sigma_r)\tilde{p}$ plane and the π-plane, with σ_a and σ_r being the axial and radial stresses in triaxial conditions. The critical state condition shown by the bold solid lines in Figure 4.2.1 is derived using $d\varepsilon_v^p = 0$ by Ohta and Nishihara (1985).

If a new stress ratio η_k is defined here,

$$\eta_k = \frac{3}{2\eta^*}(\eta_{ij} - \eta_{ij0})\eta_{ij} \tag{4.2.4}$$

The critical state condition can then be expressed as

$$\eta_k = M \tag{4.2.5}$$

Differentiating Eq. (4.2.1) gives the following stress–dilatancy relation of the Sekiguchi-Ohta model

$$\eta_k = M - \frac{d\varepsilon_v^p}{d\varepsilon_d^p} \tag{4.2.6}$$

where ε_d^p is the plastic deviatoric strain. Eq. (4.2.6) is similar in form to the stress–dilatancy relation of the original Cam-clay model (i.e., $\eta = q/p = M - d\varepsilon_v^p/d\varepsilon_d^p$), and can be drawn as shown in Figure 4.2.2 when $K_0 = 0.5$.

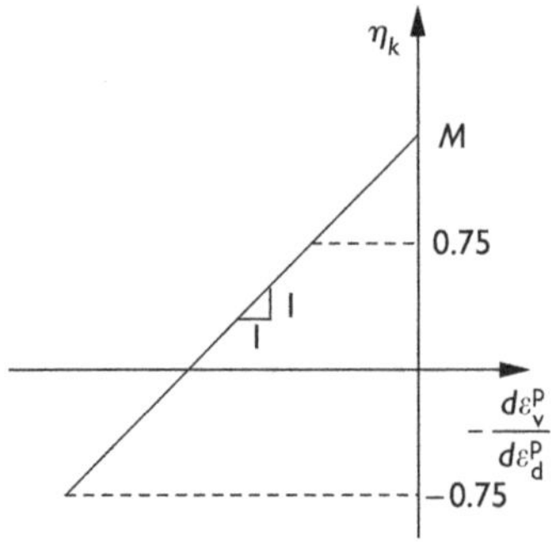

Figure 4.2.2 Stress–dilatancy relation of Sekiguchi-Ohta's model ($K_0 = 0.5$).

4.2.3 *An anisotropic hardening elastoplastic model for K_0-consolidated clays and sands*

As explained before, the Extended Mises criterion with the initial stress anisotropy ($\eta^* =$ constant) is adopted as the criterion for shear yielding in the Sekiguchi-Ohta model. As is well known, the Extended Mises criterion is not appropriated for describing soil mechanical behavior and the SMP criterion is one of the best criteria for describing shear yield and failure behavior of soils in 3D stress states. The SMP criterion has been successfully applied to the Cam-clay models by using a transformed stress $\tilde{\sigma}_{ij}$ instead of the conventional stress σ_{ij}, as described in Chapter 3. Here, the Sekiguchi-Ohta model is generalized using the transformed stress $\tilde{\sigma}_{ij}$ based on the SMP criterion.

If we replace the stress tensor σ_{ij} by the transformed stress tensor $\tilde{\sigma}_{ij}$, and introduce a new hardening parameter H instead of the original hardening parameter ε_v^p to the Sekiguchi-Ohta model, we can obtain the following yield function (f) and plastic potential function (g):

$$f = g = \frac{\lambda - \kappa}{1 + e_0}\left(\ln\frac{\tilde{p}}{\tilde{p}_0} + \frac{\tilde{\eta}^*}{M}\right) - H = 0 \tag{4.2.7}$$

where $\tilde{p}_0$ is the value of $\tilde{p}(= \tilde{\sigma}_{ii}/3)$ at the initial stress state with the void ratio e_0. For dilatant soils such as medium to dense sand, M is the value of q/p at the phase transformation where the rate of volume change is zero, i.e., $d\varepsilon_v = 0$. For non-dilatant soil such as normally consolidated clay, M is the value of q/p at the critical state in triaxial compression stress. Although the value of M for sand varies with the confining pressure, M is usually assumed to be constant for the sake of simplicity and practical application. Moreover, a new stress ratio $\tilde{\eta}^*$ in the same form as η^* defined in Eq. (4.2.2) is introduced

$$\tilde{\eta}^* = \sqrt{\frac{3}{2}(\tilde{\eta}_{ij} - \tilde{\eta}_{ij0})(\tilde{\eta}_{ij} - \tilde{\eta}_{ij0})} \tag{4.2.8}$$

in which

$$\tilde{\eta}_{ij} = \frac{(\tilde{\sigma}_{ij} - \tilde{p}\delta_{ij})}{\tilde{p}} \tag{4.2.9}$$

$\tilde{\eta}_{ij0}$ is the value of $\tilde{\eta}_{ij}$ at the initial stress state. It can be seen from Eqs (4.2.8) and (4.2.9) that the proposed yield surface is anisotropic because the tensor of the initial stress ratio $\tilde{\eta}_{ij0}$ is fixed.

Following the isotropic hardening model proposed by Yao *et al.* (1999), the increment of the hardening parameter $\tilde{H}$ for anisotropic hardening is defined as

$$d\tilde{H} = \frac{M^4}{M_f^4}\frac{M_f^4 - \tilde{\eta}_k^4}{M^4 - \tilde{\eta}_k^4} d\varepsilon_v^p \tag{4.2.10}$$

where M_f is the value of q/p at the failure state ($\varepsilon_d^p \to \infty$) in triaxial compression stress, and is also assumed to be constant for the sake of simplicity and practical application. $\tilde{\eta}_k$ is defined in the same form as Eq. (4.2.4)

$$\tilde{\eta}_k = \frac{3}{2\tilde{\eta}^*}(\tilde{\eta}_{ij} - \tilde{\eta}_{ij0})\tilde{\eta}_{ij} \tag{4.2.11}$$

By using $\tilde{\eta}_k$, we can obtain the following stress–dilatancy relation in the new model, which is in the same form as that of the Sekiguchi-Ohta model, i.e.,

$$\tilde{\eta}_k = M - \frac{d\varepsilon_v^p}{d\varepsilon_d^p} \tag{4.2.12}$$

Figure 4.2.3 shows the shape of Eq. (4.2.12) when $K_0 = 0.5$.

Taking Eq. (4.2.12) into consideration, the hardening parameter $\tilde{H}$ defined in Eq. (4.2.10) is a combination of both plastic volumetric strain ε_v^p and plastic shear strain ε_d^p, which is the same concept first introduced by Nova and Wood (1979).

Figure 4.2.4 shows the present yield curves (Eq. (4.2.7), fine solid curves), failure state lines ($\tilde{\eta}_k = M_f$, bold solid lines), and phase transformation lines ($\tilde{\eta}_k = M$, dotted line) in triaxial compression and extension stresses. From Eq. (4.2.10), it is known that, when $M_f = M$, the failure state line becomes the phase transformation line, and the hardening parameter $\tilde{H}$ becomes the plastic volumetric strain ε_v^p, which is widely used as the hardening parameter in the elastoplastic model such as the Cam-clay model and the Sekiguchi-Ohta model for normally consolidated clay.

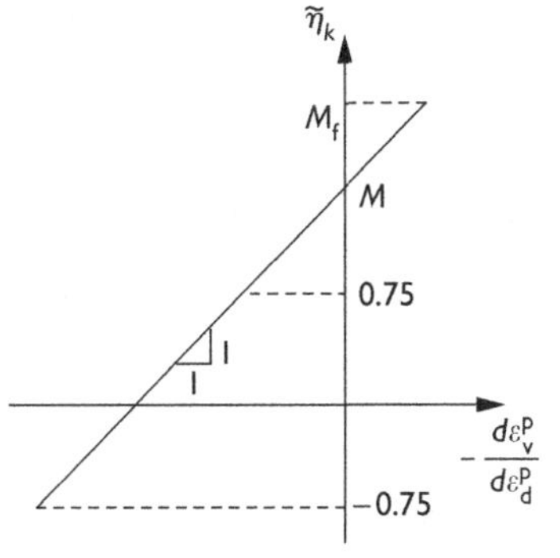

Figure 4.2.3 New stress–dilatancy relationship ($K_0 = 0.5$).

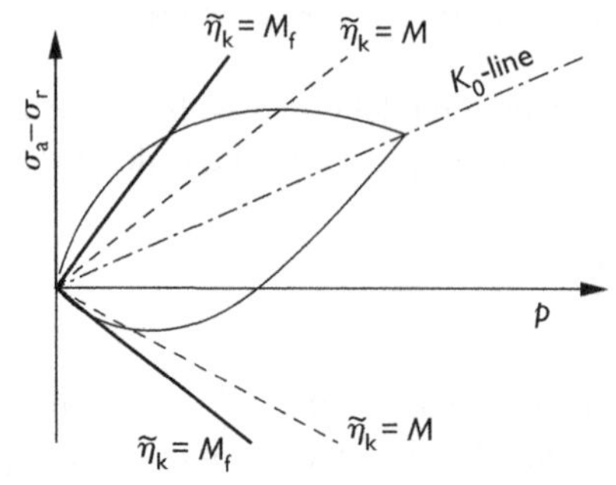

Figure 4.2.4 Proposed yield curve (fine solid line), failure state lines (bold solid line), and phase transformation lines (dotted line) in triaxial stress.

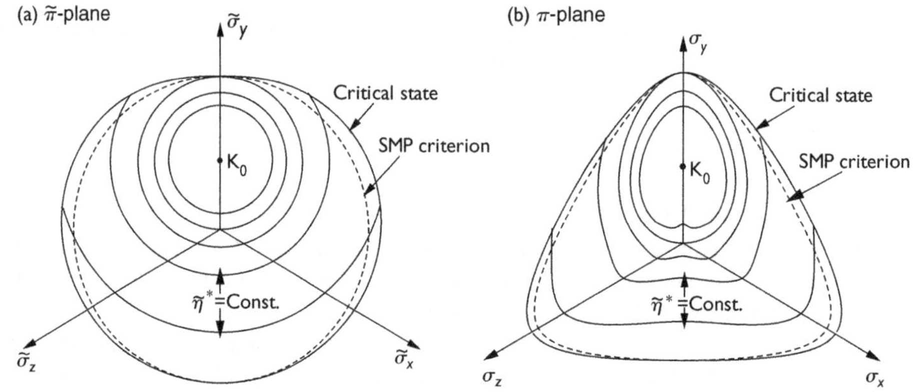

Figure 4.2.5 Yield loci of the present model in deviatoric plane.

Figure 4.2.5 shows the yield loci of the present model in the deviatoric plane. The yield loci in the transformed π-plane ($\tilde{\pi}$-plane) are concentric circles with the center being the point K_0 as shown in Figure 4.2.5(a), which are in the same shape as those of the Sekiguchi-Ohta model in the π-plane, while the yield loci become oval shaped in the ordinary π-plane as shown in Figure 4.2.5(b). It is interesting to note that in the π-plane the shape of the yield loci (Fig. 4.2.5(b)) is similar to the observed results obtained by Yamada (1979) using a true triaxial apparatus, as shown in Figure 4.2.6. The method for determining the yield point by Yamada (1979) was as follows. All specimens were first subjected to a deviatoric stress along y-direction (i.e. $\theta = 0°$) from O to A under constant mean stress. This stress path corresponds to that in a conventional triaxial compression test. After completion of the first cycle the specimens were subjected to a loading along the stress path of $\theta = 15°$, or $30°$, or $45°$ to determine the yield stress points marked with ○ from the points where the curvature of the shear stress versus shear strain curve is maximum. The method for determining the yield points marked with • are similar, but the deviatoric stress is larger. So the two yield loci shown in Figure 4.2.6 are the observed results of sand samples subjected to two different values of deviatoric stress in the y-direction.

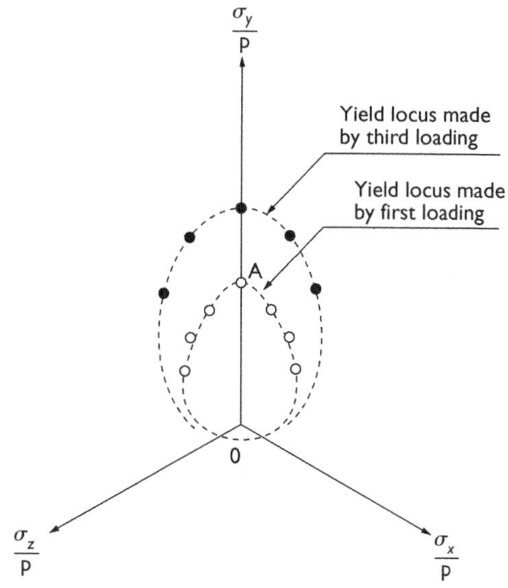

Figure 4.2.6 Yield loci in π-plane determined by undrained cyclic loading true triaxial tests on sand (after Yamada 1979).

In the present model, the associated flow rule is adopted in the transformed stress $\tilde{\sigma}_{ij}$-space, i.e.,

$$d\varepsilon_{ij}^{p} = \Lambda \frac{\partial f}{\partial \tilde{\sigma}_{ij}} \tag{4.2.13}$$

where Λ is the proportionality constant, and can be determined from the consistency condition as follows:

$$df = \frac{\partial f}{\partial p} d\tilde{p} + \frac{\partial f}{\partial \eta^*} d\tilde{\eta}^* + \frac{\partial f}{\partial H} d\tilde{H} = 0 \tag{4.2.14}$$

Substituting Eqs (4.2.10) and (4.2.13) into Eq. (4.2.14) gives

$$\frac{\partial f}{\partial \tilde{p}} d\tilde{p} + \frac{\partial f}{\partial \tilde{\eta}^*} d\tilde{\eta}^* + \frac{\partial f}{\partial \tilde{H}} \frac{M^4(M_f^4 - \tilde{\eta}_k^4)}{M_f^4(M^4 - \tilde{\eta}_k^4)} \Lambda \frac{\partial f}{\partial \tilde{\sigma}_{ii}} = 0 \tag{4.2.15}$$

Hence,

$$\Lambda = -\frac{M_f^4(M^4 - \tilde{\eta}_k^4)}{M^4(M_f^4 - \tilde{\eta}_k^4)} \frac{(\partial f/\partial \tilde{p})d\tilde{p} + (\partial f/\partial \tilde{\eta}^*)d\tilde{\eta}^*}{(\partial f/\partial \tilde{H})(\partial f/\partial \tilde{\sigma}_{ii})} \tag{4.2.16}$$

in which,

$$\frac{\partial f}{\partial \tilde{p}} = \frac{\lambda - \kappa}{1 + e_0} \frac{1}{\tilde{p}} \tag{4.2.17}$$

$$\frac{\partial f}{\partial \tilde{\eta}^*} = \frac{\lambda - \kappa}{1 + e_0} \frac{1}{M} \tag{4.2.18}$$

$$\frac{\partial f}{\partial \tilde{H}} = -1 \tag{4.2.19}$$

$$\frac{\partial f}{\partial \tilde{\sigma}_{ij}} = \frac{\partial f}{\partial \tilde{p}} \frac{\partial \tilde{p}}{\partial \tilde{\sigma}_{ij}} + \frac{\partial f}{\partial \tilde{\eta}^*} \frac{\partial \tilde{\eta}^*}{\partial \tilde{\sigma}_{ij}} \tag{4.2.20}$$

$$\frac{\partial \tilde{p}}{\partial \tilde{\sigma}_{ij}} = \frac{1}{3}\delta_{ij} \tag{4.2.21}$$

$$\frac{\partial \tilde{\eta}^*}{\partial \tilde{\sigma}_{ij}} = \frac{1}{2\tilde{\eta}^* \tilde{p}}\{3(\tilde{\eta}_{ij} - \tilde{\eta}_{ij0}) - \tilde{\eta}_{kl}(\tilde{\eta}_{kl} - \tilde{\eta}_{kl0})\delta_{ij}\} \tag{4.2.22}$$

From Eqs (4.2.17)–(4.2.22), we have

$$\frac{\partial f}{\partial \tilde{\sigma}_{ii}} = \frac{\lambda - \kappa}{1 + e_0} \frac{(M - \tilde{\eta}_k)}{M\tilde{p}} \tag{4.2.23}$$

Substituting Eqs (4.2.19) and (4.2.23) into Eq. (4.2.16) gives

$$\Lambda = \frac{1 + e_0}{\lambda - \kappa} \frac{M_{\mathrm{f}}^4 \left(M^2 + \tilde{\eta}_k^2\right)(M + \tilde{\eta}_k)\tilde{p}}{M^3 \left(M_{\mathrm{f}}^4 - \tilde{\eta}_k^4\right)} \left(\frac{\partial f}{\partial \tilde{p}} d\tilde{p} + \frac{\partial f}{\partial \tilde{\eta}^*} d\tilde{\eta}^*\right) \tag{4.2.24}$$

It can be seen from Eq. (4.2.24) that the condition $\Lambda > 0$ is satisfied during yielding $(\partial f/\partial \tilde{\sigma}_{ij} d\tilde{\sigma}_{ij} = \partial f/\partial \tilde{p} d\tilde{p} + \partial f/\partial \tilde{\eta}^* d\tilde{\eta}^* > 0)$.

From $d\tilde{H} = 0$ and $\varepsilon_{\mathrm{d}}^{\mathrm{p}} \to \infty$ and Eqs (4.2.10) and (4.2.12), we can obtain the failure criterion of the present model as follows:

$$\tilde{\eta}_k = M_{\mathrm{f}} \tag{4.2.25}$$

4.2.4 Comparison of model predictions with experimental results

The model contains six material parameters λ, κ, M, M_{f}, ν, and e_0. In comparison with the Sekiguchi-Ohta model, the present model uses one more parameter, i.e., M_{f}. However, for the soils with only negative dilatancy, the material parameters of two models are identical, because $M_{\mathrm{f}} = M$ in this case.

4.2.4.1 Modeling triaxial behavior

Figure 4.2.7 compares the predicted results obtained from the Sekiguchi-Ohta model (abbreviated as S-O model in the figure; denoted by the dotted lines) and the present model (solid lines) with experimental results of triaxial tests (sign ○) on a non-dilatant soil (experimental data from Ichihara (1997)). The soil is a mixture of Toyoura sand and Fujinomori clay. The mixture of Toyoura sand to Fujinomori clay by dry weight is 7 : 3. In that figure, ε_{a}

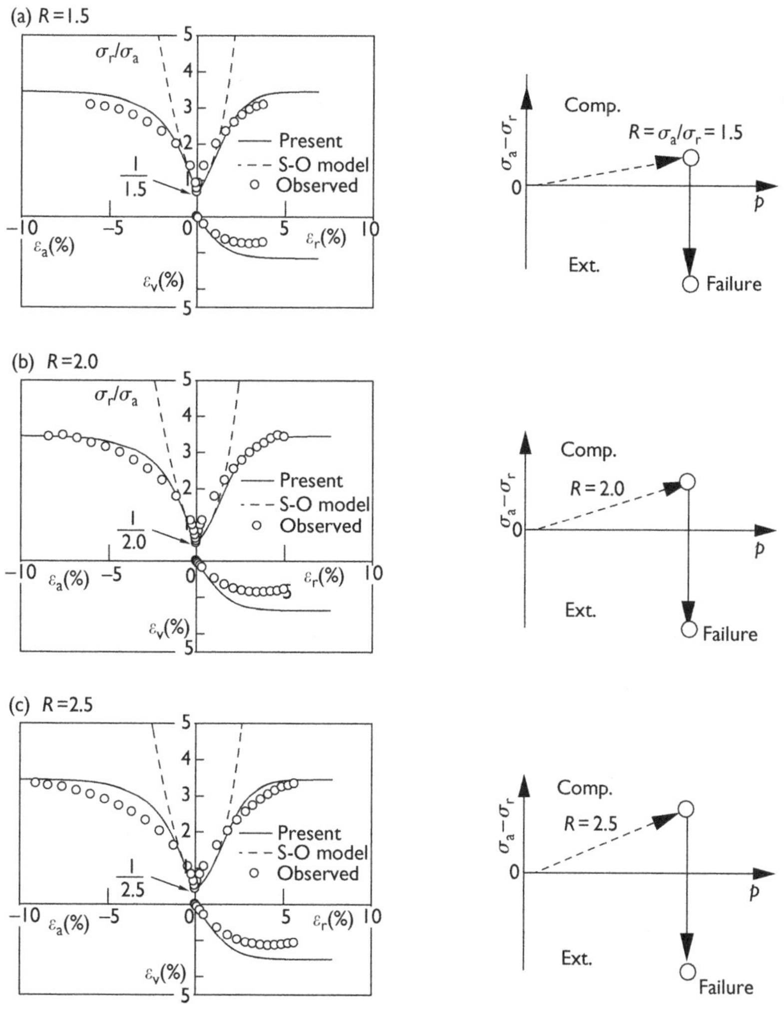

Figure 4.2.7 Comparison between predicted and measured stress–strain behavior during shear from anisotropic stresses.

and ε_r are the axial and radial strains in triaxial tests. The material parameters used in the prediction are as follows: $\lambda/(1+e_0) = 0.021$, $\kappa/(1+e_0) = 0.0034$, $M_f = M = 1.32$, and $\nu = 0.3$. The triaxial specimens were first anisotropically consolidated up to $p = 980$ kPa along the stress paths of $\sigma_a/\sigma_r = 1.5$, 2.0, and 2.5 respectively in triaxial compression, and then sheared to failure in triaxial extension under constant p, as shown in the right-hand side of Figure 4.2.7. The left-hand side of Figure 4.2.7 shows the predictions and triaxial test results during shearing. It can be seen that the present model better predicts the experimental results than the Sekiguchi-Ohta model.

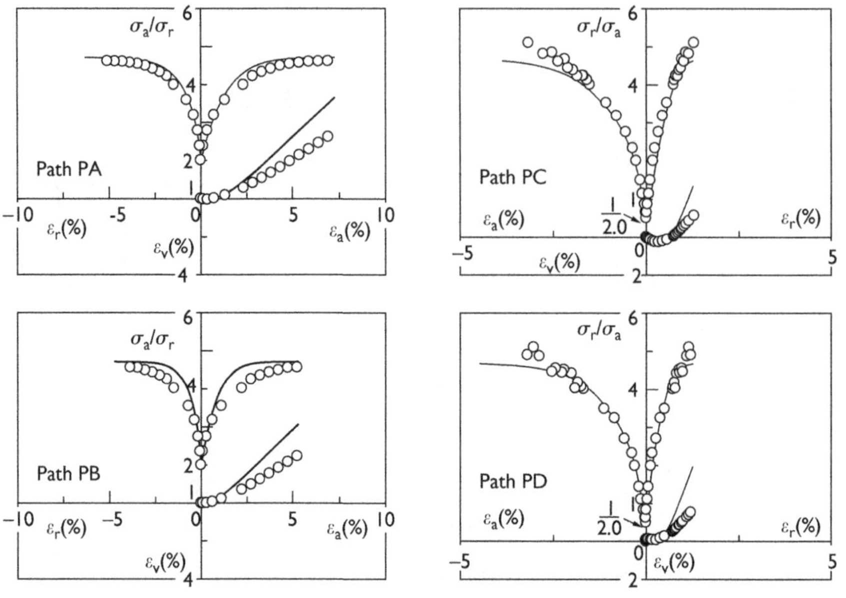

Figure 4.2.8 Comparison between predicted and measured stress–strain behavior of sand shearing from anisotropic stress.

Figure 4.2.8 shows the experimental and predicted results of triaxial compression and extension tests on Toyoura sand, starting with $K_0(= 0.5)$-consolidation state (experimental data from Funada *et al.* (1989)). The stress paths tested are from point P at $\sigma_a/\sigma_r = 2$ in triaxial compression to triaxial compression or triaxial extension up to failure under $p = 196$ kPa or $\sigma_r = 147$ kPa, as shown in Figure 4.2.9. The material parameters are $\lambda/(1+e_0) = 0.00403$, $\kappa/(1+e_0) = 0.00251$, $M = 0.95$, $M_f = 1.66$, and $\nu = 0.3$. The model can properly predict the sand behavior including dilatancy characteristics.

It can be seen from Figures 4.2.7 and 4.2.8 that the present model can well predict the stress–strain relation of clays and sands during shearing from initially anisotropic stress states to triaxial compression or triaxial extension stress states. Therefore, it can be deduced that the present model is capable of predicting the stress–strain relations of non-dilatant and dilatant soils during shearing from an anisotropic stress state to any other stress states in 3D stress state.

4.2.4.2 Modeling plane strain behavior

In numerical analysis of geotechnical structures like strip footing, retaining walls and embankments, plane strain condition is usually assumed. In order to validate the present model under plane strain condition, the model is used to predict the experimental data of plane strain on K_0-consolidated Fujinomori clay (Nakai *et al.* 1986, 1987).

Figure 4.2.10 shows the stress paths tested in the plane strain tests. The tests are conducted along six kinds of stress paths (AB, AC, AD, AB′, AC′, and AD′) under plane strain condition from K_0-consolidation state (point A: $\sigma_y = 196$ kPa, $\sigma_x = \sigma_z = 98$ kPa). The model

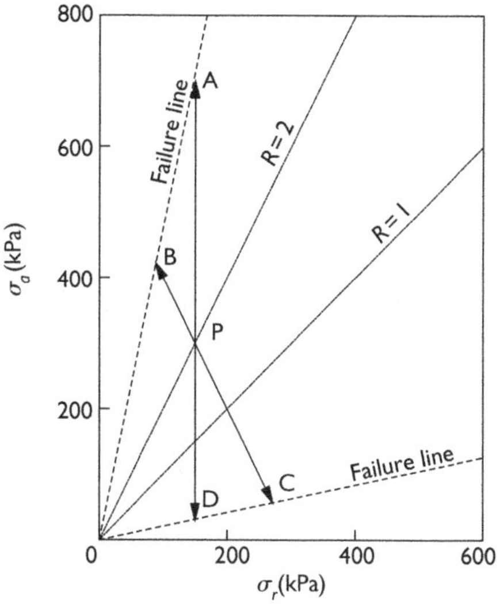

Figure 4.2.9 Stress paths for triaxial tests on sand.

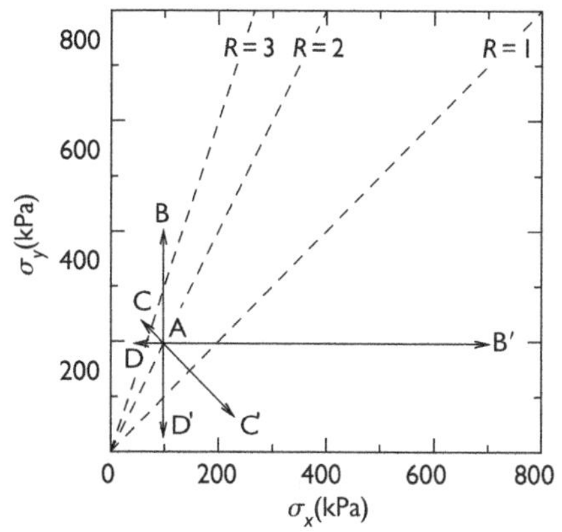

Figure 4.2.10 Stress paths under plane strain condition.

parameters used in the predictions are either from literature (Nakai *et al.* 1986, 1987) or determined from the results of triaxial tests and isotropic compression and swelling tests.

Figure 4.2.11 shows the comparison between the measured intermediate principal stress and the predicted results obtained using the present model and the Sekiguchi-Ohta model. Because the Sekiguchi-Ohta model uses the Extended Mises criterion as the failure criterion, the values of σ_y/σ_x and σ_z/σ_x at failure are larger than the measured values. The present model gives a relatively better predict of the measured values.

Figure 4.2.12 shows the measured stress states along the stress paths AC and AC′ in the π-plane, and the corresponding results predicted by the proposed model and the Sekiguchi-Ohta model. It can be seen that the present model predicts the stress paths in the π-plane

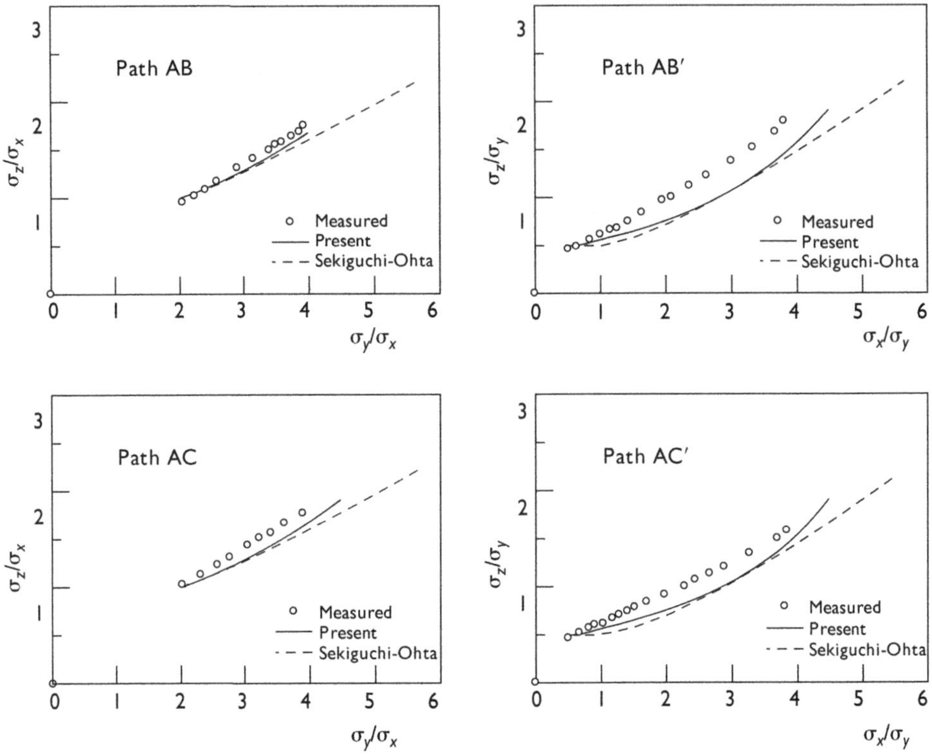

Figure 4.2.11 Comparison between predicted and measured intermediate principal stress under plane strain condition (data after Nakai *et al.* 1986).

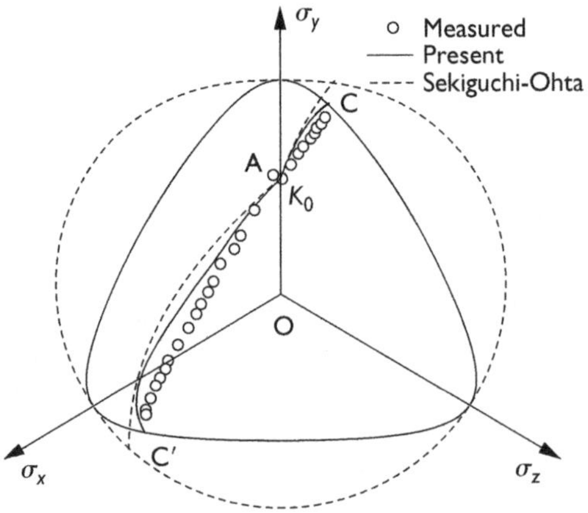

Figure 4.2.12 Predicted and measured stress paths under plane strain in the π-plane.

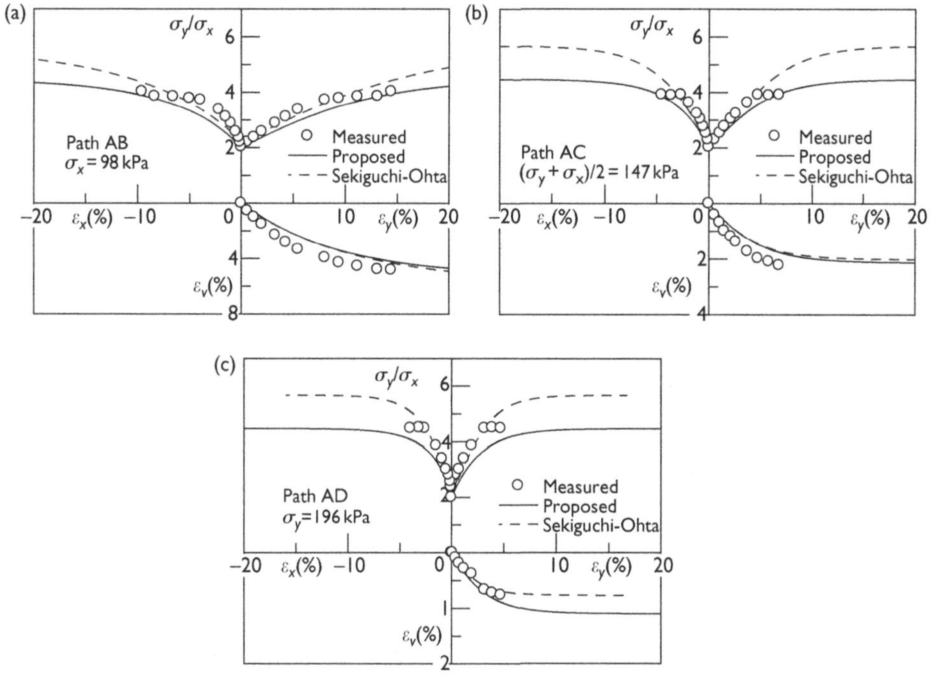

Figure 4.2.13 Comparison between predicted and measured stress–strain behavior under plane strain condition without change in principal stress direction (data after Nakai *et al.* 1986).

better than the Sekiguchi-Ohta model does. The critical states (Points C and C′) obtained using the present model are close to the measured stress states at failure, while the stress states obtained using the Sekiguchi-Ohta model are larger than the measured values.

Figures 4.2.13 and 4.2.14 show the experimental and predicted results of the plane strain tests along the stress paths shown in Figure 4.2.10, in relation between the principal stress ratios (σ_y/σ_x and σ_x/σ_y), principal strains (ε_x and ε_y) and the volumetric strain (ε_v).

It is seen that both the models can predict, at least to some extent, the stress–strain response under plane strain condition. This is because the SMP criterion and the Mises criterion are relatively close under plane strain condition (see Fig. 4.2.12). However, the results predicted by the present model better fit the experimental data from the six cases than those by the Sekiguchi-Ohta model, compared with the measured results of the six cases.

It can be concluded from the results shown in Figures 4.2.11–4.12.14 that the Sekiguchi-Ohta model can give acceptable predictions of tests results for K_0-consolidated clay, except for the strength under plane strain condition. This is the reason why the Sekiguchi-Ohta model can predict, to some extent, deformation of earth structures under plane strain condition.

4.2.5 *Elastoplastic constitutive tensor*

To derive the elastoplastic constitutive tensor for the present model to solve boundary value problems by means of the finite element method, we need to replace Eq. (A3.2.10) in

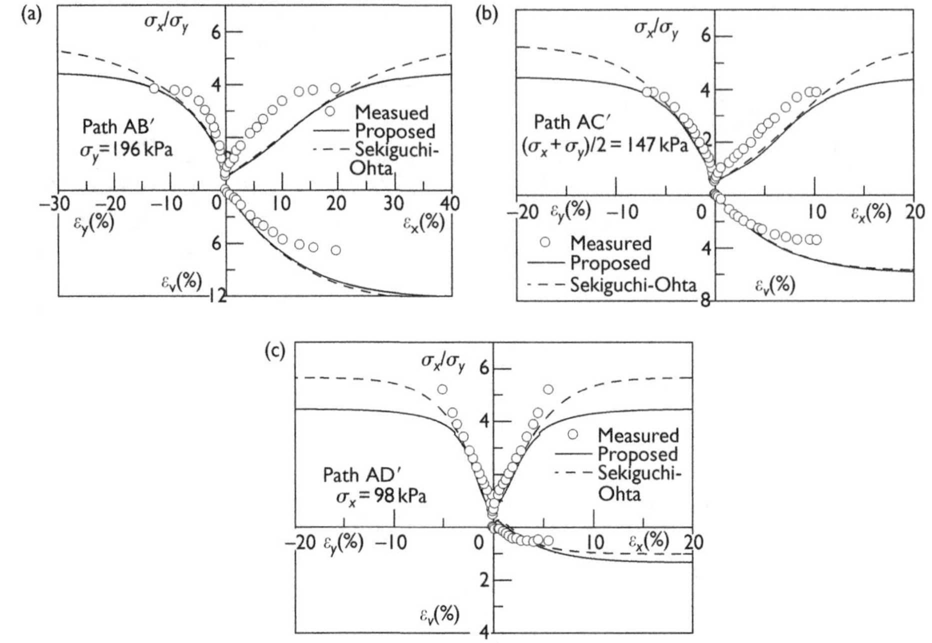

Figure 4.2.14 Comparison between predicted and measured stress–strain behavior under plane strain condition with change in principal stress directions (data after Nakai *et al.* 1986).

Chapter 3 with the following equation

$$X = \frac{M_f^4 \left(M^4 - \tilde{\eta}_k^4\right)}{M^4 \left(M_f^4 - \tilde{\eta}_k^4\right)} \frac{\partial f}{\partial \tilde{\sigma}_{ii}} + \frac{\partial f}{\partial \sigma_{ij}} D_{ijkl}^{e} \frac{\partial f}{\partial \tilde{\sigma}_{kl}} \tag{4.2.26}$$

Substituting Eq. (4.2.23) into Eq. (4.2.26) gives

$$X = \frac{\lambda - \kappa}{1 + e_0} \frac{M^3 \left(M_f^4 - \tilde{\eta}_1^4\right)}{M_f^4 \left(M^2 + \tilde{\eta}_1^2\right) \left(M + \tilde{\eta}_1\right) \tilde{p}} + \frac{\partial f}{\partial \sigma_{ij}} D_{ijkl}^{e} \frac{\partial f}{\partial \tilde{\sigma}_{kl}} \tag{4.2.27}$$

4.2.6 Conclusions

This section presented an anisotropic hardening model for K_0-consolidated clays, silts, and sands within a simple framework. Several conclusions can be drawn.

1 The yield locus for soils subjected to the anisotropic stress should be oval shaped in the π-plane.
2 A unified and simple elastoplastic model for K_0-consolidated clays and sands was proposed, which is capable of describing deformation and strength characteristics of clays and sands with initially anisotropic stress states in 3D space.

3 For non-dilatant soil, the model parameters are the same as those of the original Cam-clay model or the Sekiguchi-Ohta model. For dilatant soil, only one additional parameter is required for the present model.

4.3 An elastoplastic model for unsaturated soils

4.3.1 *Introduction*

Figure 4.3.1 shows a two-dimensional model of granular material under unsaturated condition. Soil particles are modeled by a stack of aluminum rods with diameters of 1.6 and 3 mm. It can be seen that some water is attached to the aluminum rods (soil particles) due to surface tension and thus capillary menisci form between particles in unsaturated soils. The pressure difference between pore air and pore water inside menisci is called suction s, i.e., $s = u_a - u_w$. The curved air-water interfaces cause pore-water tension, which in turn generates inter-particle forces F, as shown in Figure 4.3.2. The key point for modeling unsaturated soil behavior using plasticity theory is thus how to account for the effect of the suction s or the inter-particle force F on the effective stress, strength, yield stress, and hardening of unsaturated soils.

In the following sections, we will describe the effective stress and failure criterion for unsaturated soils, and modeling stress–strain relationships under isotropic and general stress states. The model predictions are then compared with the experimental results obtained from isotropic compression, triaxial compression and extension tests under constant suction and decreasing suction (Sun *et al.* 2000).

Figure 4.3.1 A picture of water attachment in an assembly of wet cylindrical aluminum rods.

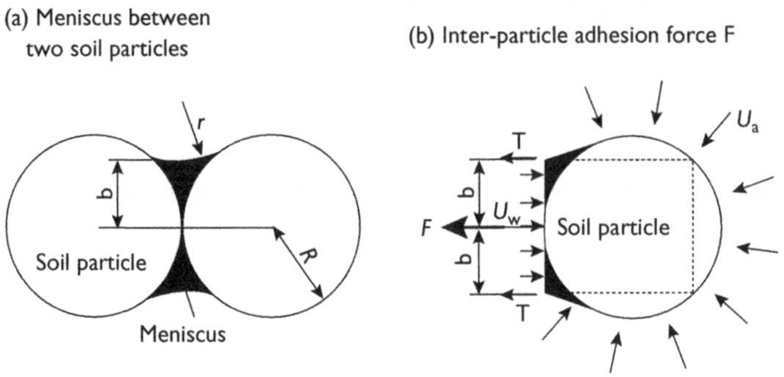

Figure 4.3.2 Inter-particle adhesive force due to water meniscus.

The following symbols are used for stress and pressure.

σ_{tij}: total stress tensor
s: suction ($= u_a - u_w$; u_a: pore-air pressure, u_w: pore-water pressure)
σ_{ij}: net stress tensor ($= \sigma_{tij} - u_a\delta_{ij}$; δ_{ij}: Kronecker's delta)
p : mean net stress ($= \sigma_{ii}/3$)
q: deviatoric stress ($= \sqrt{3(\sigma_{ij} - p\delta_{ij})(\sigma_{ij} - p\delta_{ij})/2}$)

4.3.2 *Effective stress for unsaturated soils*

As mentioned in the previous section, an unsaturated soil can be considered as a kind of cohesive-frictional material (briefly known as $c - \phi$ material) because of the inter-particle forces in the soil. By introducing a parameter $\sigma_0(= c \cdot \cot\phi)$ that can evaluate cohesion due to suction, a translated principal stress $\hat{\sigma}_i$ and a translated stress tensor $\hat{\sigma}_{ij}$ can be defined as follows:

$$\hat{\sigma}_i = \sigma_i + \sigma_0 \tag{4.3.1}$$

$$\hat{\sigma}_{ij} = \sigma_{ij} + \sigma_0\delta_{ij} \tag{4.3.2}$$

Bishop (1959) suggested a tentative expression for effective stress:

$$\hat{\sigma} = (\sigma_t - u_a) + \chi(u_a - u_w) \tag{4.3.3}$$

Where σ_t is a total stress, χ is a parameter related to the degree of saturation S_r of the soil. The magnitude of the χ parameter is unity for a saturated soil and zero for a dry soil. The relationship between χ and the degree of saturation S_r was obtained experimentally. Because the degree of saturation S_r is related to suction s, χ is also a function of suction.

Comparing Eq. (4.3.2) with Eq. (4.3.3), a general effective stress equation for unsaturated soils can be written by

$$\hat{\sigma}_{ij} = (\sigma_{tij} - u_a\delta_{ij}) + \sigma_0(s)\delta_{ij} \tag{4.3.4}$$

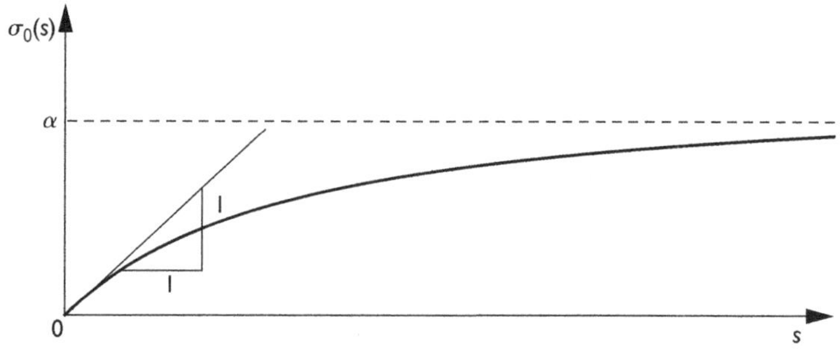

Figure 4.3.3 Adopted relationship between suction s and corresponding suction stress $\sigma_0(s)$.

where $(\sigma_{tij} - u_a\delta_{ij})$ is a net stress when the pore-air pressure u_a is acting. $\sigma_0(s)$ is a kind of bonding stress due to suction and is called suction stress here. Suction stress $\sigma_0(s)$ is assumed to be a hyperbolic function of suction (Fig. 4.3.3).

$$\sigma_0(s) = \frac{a\,s}{a+s} \tag{4.3.5}$$

where a is a material parameter, and is a value of $\sigma_0(s)$ at $s \to \infty$ (i.e., dry soil). Eq. (4.3.5) is based on the following considerations. First, the strength of unsaturated soil increases nonlinearly with suction and approaches a maximum at a high suction (Escario and Saez 1987; Escario and Juca 1989). Second, when the suction is applied to saturated soil in the initial stage, $\sigma_0(s)$ can be considered to be equal to s, because the principle of effective stress holds near saturated state.

When $i = j$ substituting Eq. (4.3.4) into Eq. (4.3.3) gives

$$\hat{\sigma} = (\sigma_t - u_a) + \frac{a}{a+s}(u_a - u_w) \tag{4.3.6}$$

From Eq. (4.3.6) in the case of saturated soil ($s = 0$),

$$\hat{\sigma} = \sigma_t - u_w \tag{4.3.7}$$

This equation is the same as the effective stress equation by Terzaghi (1936). On the other hand, for dry soil (i.e., $s \to \infty$) Eq. (4.3.6) becomes

$$\hat{\sigma} = \sigma_t - u_a + a \tag{4.3.8}$$

Comparing Eq. (4.3.3) with Eq. (4.3.6) gives

$$\chi = \frac{a}{a+s} \tag{4.3.9}$$

It is clear that $\chi = 1$ when $s = 0$ (i.e., saturated soil), and $\chi = 0$ when $s \to \infty$ (i.e., dry soil).

4.3.3 *Transformed stress tensor based on Extended SMP criterion*

As described in Section 1.5, the concept of Extended SMP has been successfully applied to some cohesive-frictional materials such as cemented soil (Matsuoka and Sun 1995). As the shear yield or failure criterion for cohesive-frictional materials, the Extended SMP criterion can be expressed as

$$\frac{\hat{I}_1 \hat{I}_2}{\hat{I}_3} = \text{constant} \tag{4.3.10}$$

where $\hat{I}_1$, $\hat{I}_2$, and $\hat{I}_3$ are the first, second, and third invariants of the translated stress tensor, which becomes the effective stress for unsaturated soils as expressed in Eq. (4.3.4).

Figure 4.3.4 shows the Extended SMP criterion in three-dimensional principal stress (σ_i or $\hat{\sigma}_i$) space. The Extended SMP criterion is reduced to the SMP criterion for frictional materials such as sands when $\sigma_0 = 0$, and the Mises criterion for cohesive materials such as metals when $\sigma_0 \to \infty$. So, it is considered that the Extended SMP criterion is more reasonable than the Extended Mises criterion with cohesion (i.e., $q/(p + \sigma_0) = $ const.), which is adopted as shear yield and failure criteria in some existing models for unsaturated soil. In order to introduce the Extended SMP criterion to the elastoplastic model for cohesive-frictional materials, a transformed stress $\tilde{\hat{\sigma}}_{ij}$ based on the Extended SMP criterion has been proposed. The transformed stress $\tilde{\hat{\sigma}}_{ij}$ is derived from the transformation of the Extended SMP criterion to a cone with the transformed principal stress ($\tilde{\hat{\sigma}}_i$) being the space diagonal in Figure 4.3.5, that is, the Extended SMP curve becomes a circle in the π-plane of the transformed principal stress ($\tilde{\hat{\sigma}}_i$) space (Fig. 4.3.6), i.e.,

$$\tilde{\hat{\sigma}}_{ij} = \hat{p}\delta_{ij} + \frac{\hat{\ell}_0}{\hat{\ell}_\theta}(\hat{\sigma}_{ij} - \hat{p}\delta_{ij}) \tag{4.3.11}$$

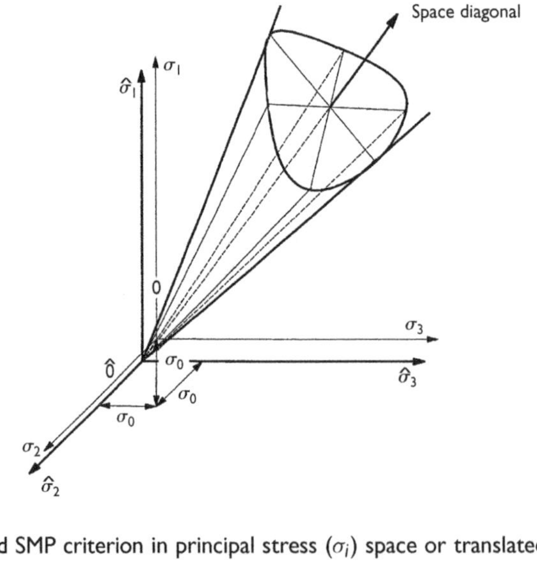

Figure 4.3.4 Extended SMP criterion in principal stress (σ_i) space or translated principal stress ($\hat{\sigma}_i$) space.

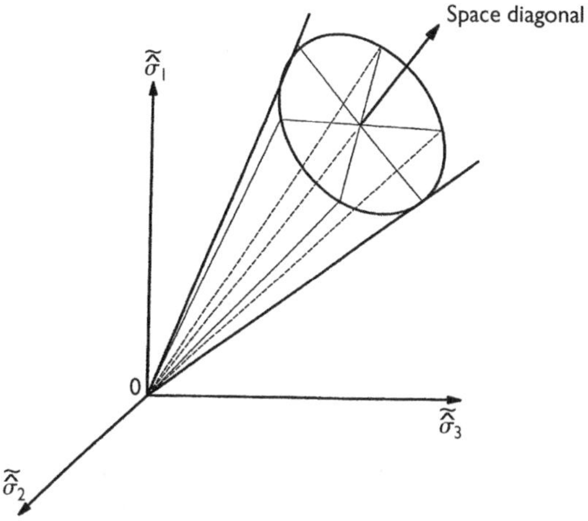

Figure 4.3.5 Extended SMP criterion in transformed principal stress ($\tilde{\hat{\sigma}}_i$) space.

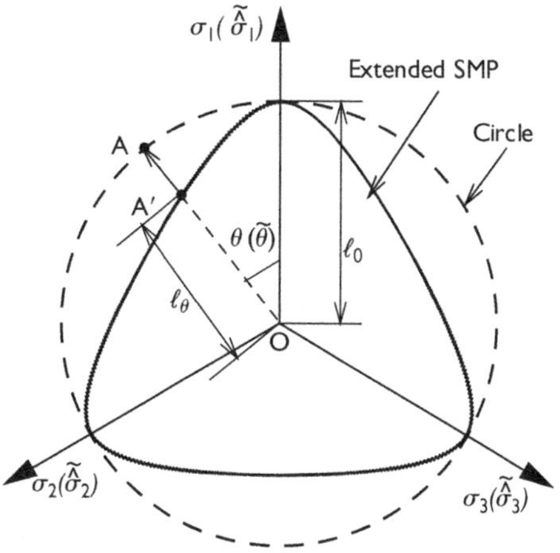

Figure 4.3.6 Extended SMP criterion in the π-plane and the transformed π-plane.

where

$$\hat{p} = \hat{\sigma}_{ii}/3 \tag{4.3.12}$$

and

$$\left.\begin{aligned} \hat{\ell}_0 &= \frac{2\sqrt{6}\hat{I}_1}{3\sqrt{(\hat{I}_1\hat{I}_2 - \hat{I}_3)/(\hat{I}_1\hat{I}_2 - 9\hat{I}_3) - 1}} \\ \hat{\ell}_\theta &= \sqrt{(\hat{\sigma}_{ij} - \hat{p}\delta_{ij})(\hat{\sigma}_{ij} - \hat{p}\delta_{ij})} \end{aligned}\right\} \tag{4.3.13}$$

$\tilde{\hat{\sigma}}_{ij}$ can be used to replace σ_{ij} for extending the existing elastoplastic models using only p and q as stress parameters to three-dimensional elastoplastic models for cohesive-frictional materials, because this modification means that the Extended SMP criterion is introduced as shear yield and failure criteria in the model.

4.3.4 *Formulation of model for unsaturated soils*

4.3.4.1 *Strength of unsaturated soil*

An extension of the general Mohr-Coulomb failure criterion to include partial saturation has been proposed by Fredlund *et al.* (1978) as follows:

$$\tau_f = c + \sigma \tan\phi_1 + s \tan\phi_2 \tag{4.3.14}$$

where c is the cohesion excluding the component induced by suction, ϕ_1 is the ordinary angle of internal friction, and ϕ_2 is a new angle indicating the rate of increase in shear strength relative to suction.

Confining this research on soils without physical cohesion such as cementation, aging effect, and electrochemical forces between soil particles, etc, i.e., $c = 0$, we can obtain the following equation from Eq. (4.3.14).

$$\tau_f = (\sigma + \sigma_0(s)) \tan\phi_1 = \hat{\sigma} \tan\phi_1 \tag{4.3.15}$$

where

$$\sigma_0(s) = s \tan\phi_2 / \tan\phi_1 \tag{4.3.16}$$

Because Eq. (4.3.14) or (4.3.15) is an extension of the general Mohr-Coulomb failure criterion, which is valid only in triaxial compression stress, a more reasonable failure criterion for unsaturated soil should be the Extended SMP failure criterion in general stresses, that is

$$\tilde{\hat{q}} = M(s)\tilde{\hat{p}} = M(s)\hat{p} = M(s)(p + \sigma_0(s)) \tag{4.3.17}$$

where $M(s)$ is the slope of $p - q$ relation in triaxial compression stress at failure, and is assumed to be linear with regard to $\sigma_0(s)$, i.e.,

$$M(s) = M(0) + M_s\sigma_0(s) \tag{4.3.18}$$

$M(0)$ is the value of $M(s)$ for $s = 0$, M_s is a material parameter. $\tilde{\hat{p}}$ and $\tilde{\hat{q}}$ are the invariants of the transformed stress tensor $\tilde{\hat{\sigma}}_{ij}$, i.e.,

$$\left.\begin{aligned} \tilde{\hat{p}} &= \tilde{\hat{\sigma}}_{ii}/3 = \hat{\sigma}_{ii}/3 \\ \tilde{\hat{q}} &= \sqrt{3(\tilde{\hat{\sigma}}_{ij} - \tilde{\hat{p}}\delta_{ij})(\tilde{\hat{\sigma}}_{ij} - \tilde{\hat{p}}\delta_{ij})/2} \end{aligned}\right\} \tag{4.3.19}$$

Eq. (4.3.17) is equal to the general Mohr-Coulomb criterion when the stress state is under triaxial compression or triaxial extension stress state, and has been verified by the results of true triaxial tests on unsaturated soil under constant suction (Matsuoka *et al.* 2002). The shape of Eq. (4.3.17) in the three-dimensional principal stress space is the same one shown in

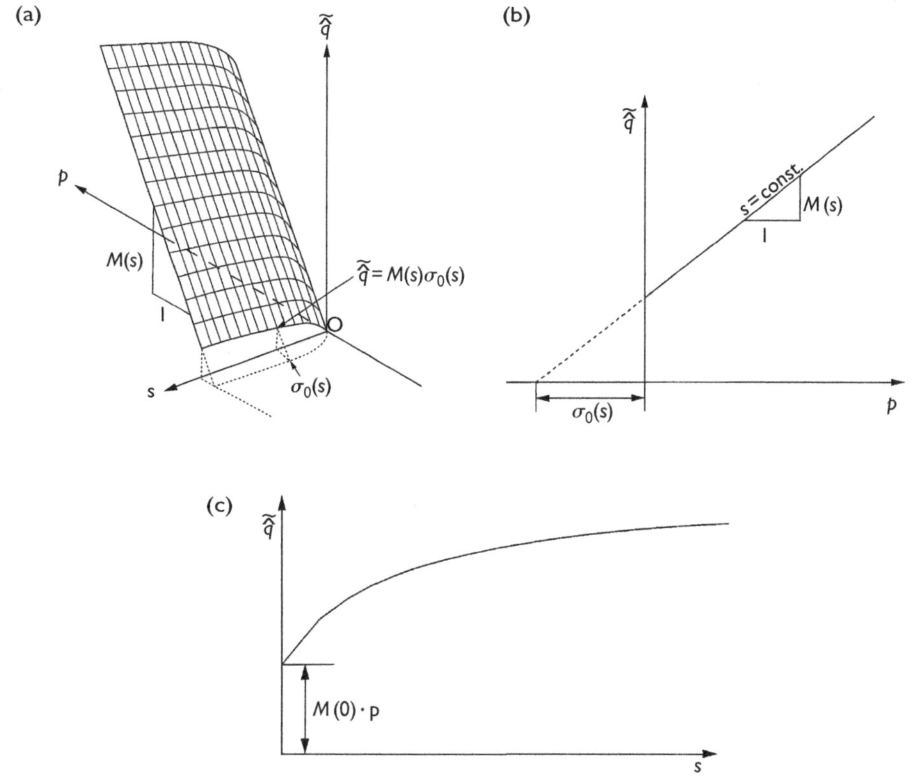

Figure 4.3.7 Failure surface of unsaturated soil under three-dimensional stress (a) in $(p, \tilde{q}, s)$ space, (b) s = constant and (c) p = constant.

Figures 4.3.4, 4.3.5, and 4.3.7(b) at a given suction. When suction changes, Eq. (4.3.17) can be drawn schematically as Figures 4.3.7(a),(c).

4.3.4.2 Formulation of model in isotropic stress state

Figure 4.3.8 shows the compression curves for saturated and unsaturated soils along the virgin isotropic loading in the $e \sim \ln \hat{p}$ plane and the $e^{\mathrm{p}} \sim \ln \hat{p}$ plane. e^{p} is the sum of the initial void ratio $e_i(s)$ and the plastic increment of void ratio (de^{p}) due to change in $\hat{p}$. The normal compression curve for saturated soil is

$$e = e_i(0) - \lambda(0) \ln \frac{p_y^*}{p_i} \tag{4.3.20}$$

and the normal compression curve for unsaturated soil with a constant suction s is

$$e = e_i(s) - \lambda(s) \ln \frac{p_y + \sigma_0(s)}{p_i} = e_i(s) - \lambda(s) \ln \frac{\hat{p}_y}{p_i} \tag{4.3.21}$$

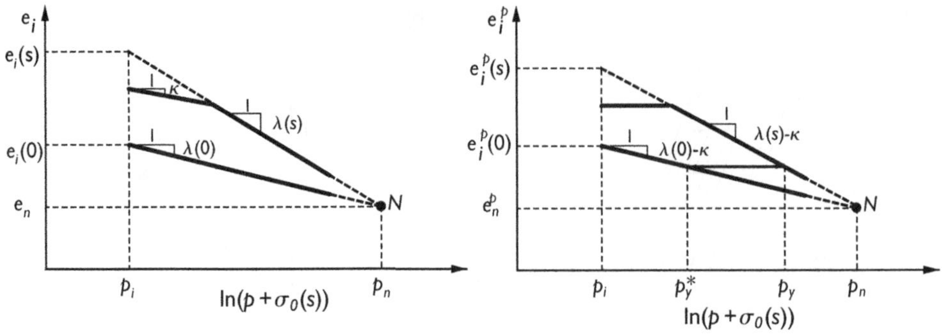

Figure 4.3.8 Relationship between compression curves for saturated and unsaturated soils.

where $\sigma_0(s)$ is a suction stress, p_i is a reference net stress, and p_y^* and p_y are the yield stresses on the normal compression curves for saturated and unsaturated soils in isotropic stress. In Figure 4.3.8(a), we assume that there exits a normal compression line between void ratio (e) and mean effective stress ($\hat{p}_y$) for a given unsaturated soil at constant suction. This assumption is similar to the normal compression line between void ratio (e) and mean effective stress (p_y^*) in the Cam-clay model for saturated clay. Here, we do not adopt a linear e $\sim \ln p$ relationship but adopt the linear e $\sim \ln \hat{p}$ relationship because $\hat{p}$ is considered to be the mean effective stress for unsaturated soils.

The point N in Figure 4.3.8(a) is considered to be unchanged when suction changes, which means that no collapse or swelling occurs at the state of point N, i.e., there exits a single point N that all the normal compression lines pass through. From the given two lines where suction are zero and s the coordinates of point N can be obtained as follows:

$$\left.\begin{aligned} p_n &= p_i e^{(e_i(0)-e_i(s))/(\lambda(0)-\lambda(s))} \\ e_n &= \frac{1}{\lambda(0)-\lambda(s)}(e_i(s)\lambda(0) - e_i(0)\lambda(s)) \end{aligned}\right\} \tag{4.3.22}$$

Eq. (4.3.21) can also be rewritten using the coordinates of point N as

$$e = e_n - \lambda(s)\ln\frac{p_y + \sigma_0(s)}{p_n} \tag{4.3.23}$$

When p_y or/and $\sigma_0(s)$ change in the normal compression region, the increment of void ratio can be obtained by differentiating Eq. (4.3.23).

$$\begin{aligned} de &= \frac{\partial e}{\partial p_y}dp_y + \frac{\partial e}{\partial \sigma_0(s)}d\sigma_0(s) \\ &= \frac{-\lambda(s)}{p_y+\sigma_0(s)}dp_y + \left(\lambda_s \ln\frac{p_n}{p_y+\sigma_0(s)} - \frac{\lambda(s)}{p_y+\sigma_0(s)}\right)d\sigma_0(s) \end{aligned} \tag{4.3.24}$$

in which $\lambda(s)$ is assumed to be linear with regard to $\sigma_0(s)$, i.e.,

$$\lambda_s = \frac{d\lambda(s)}{d\sigma_0(s)} = \frac{\lambda(s) - \lambda(0)}{\sigma_0(s)} \tag{4.3.25}$$

so,

$$\lambda(s) = \lambda(0) + \lambda_s \sigma_0(s) \tag{4.3.26}$$

As shown in Figure 4.3.8(b), from the same plastic volumetric strain of saturated and unsaturated soils, the following equation can be obtained with the assumption that the swelling indexes κ of saturated and unsaturated soils are the same.

$$(\lambda(s) - \kappa) \ln \frac{p_n}{p_y + \sigma_0(s)} = (\lambda(0) - \kappa) \ln \frac{p_n}{p_y^*} \tag{4.3.27}$$

Substituting Eq. (4.3.26) into Eq. (4.3.27) and solving for p_y gives

$$p_y = p_n \left(\frac{p_y^*}{p_n} \right)^{(\lambda(0)-\kappa)/(\lambda(0)-\kappa+\lambda_s\sigma_0(s))} - \sigma_0(s) \tag{4.3.28}$$

The contour lines of the equivalent plastic volumetric strain can be drawn as shown in Figure 4.3.9 in the $p \sim s$ plane using Eq. (4.3.28) and Eq. (4.3.5). Figure 4.3.10 show the relationship among dp_y, $d\sigma_0(s)$, and dp_y^*, which can be obtained by differentiating Eq. (4.3.28) as follows:

$$dp_y = \frac{\partial p_y}{\partial p_y^*} dp_y^* + \frac{\partial p_y}{\partial \sigma_0(s)} d\sigma_0(s) \tag{4.3.29}$$

in which

$$\frac{\partial p_y}{\partial p_y^*} = \frac{\lambda(0) - \kappa}{\lambda(0) - \kappa + \lambda_s \sigma_0(s)} \left(\frac{p_y^*}{p_n} \right)^{(-\lambda_s\sigma_0(s))/(\lambda(0)-\kappa+\lambda_s\sigma_0(s))} \tag{4.3.30}$$

$$\frac{\partial p_y}{\partial \upsilon_0(s)} = \frac{(p_y + \sigma_0(s))(\lambda(0) - \kappa)\lambda_s}{(\lambda(0) - \kappa + \lambda_s \sigma_0(s))^2} \ln \left(\frac{p_n}{p_y^*} \right) - 1 \tag{4.3.31}$$

It can be seen from Figures 4.3.9 and 4.3.10 that the increase in the plastic volumetric strain of unsaturated soil is caused by the increase in the yield stress and/or the decrease in the suction, while the increase in the plastic volumetric strain of saturated soil is caused only by the increase in the yield stress.

4.3.4.3 Formulation of model in general stress

As known, the modified Cam-clay model was well established and applied for saturated soils. Reasonably good predictions by the model have extensively been exercised for many years. We adopted this model for simplicity and clarity of physical meanings of the model parameters. Experiments on unsaturated soils have validated the Extended SMP criterion

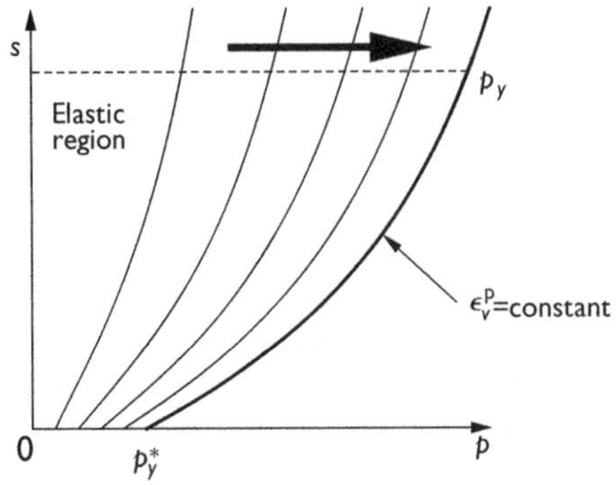

Figure 4.3.9 Contour lines of equivalent plastic volumetric strain.

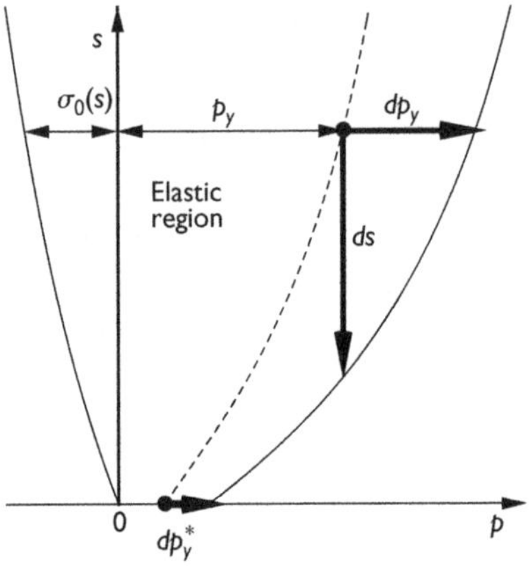

Figure 4.3.10 Relationships between dp_y, dp_y^*, and ds in isotropic stress.

for the shear strength of unsaturated soils in 3D stresses (Matsuoka *et al.* 2002). The yield function (f) and the plastic potential function (g) are proposed to have the following form:

$$f = g = \tilde{\hat{q}}^2 + M^2\tilde{\hat{p}}(\tilde{\hat{p}} - p_y - \sigma_0(s)) = 0 \tag{4.3.32}$$

Figure 4.3.11 shows the geometrical shape of Eq. (4.3.32) when the suction is at a given value for unsaturated soil and zero for saturated soil. When the suction changes, the yield surface can be drawn as in Figure 4.3.12 by allowing for Eqs (4.3.5) and (4.3.28), in which the relationships between s and $\sigma_0(s)$ and between p_y and p_y^* are given.

Because the plastic strain increment $d\varepsilon_{ij}^{\mathrm{p}}$ obeys the associated flow rule in the $\tilde{\hat{\sigma}}_{ij}$-space,

$$d\varepsilon_{ij}^{\mathrm{p}} = \Lambda \frac{\partial f}{\partial \tilde{\hat{\sigma}}_{ij}} \tag{4.3.33}$$

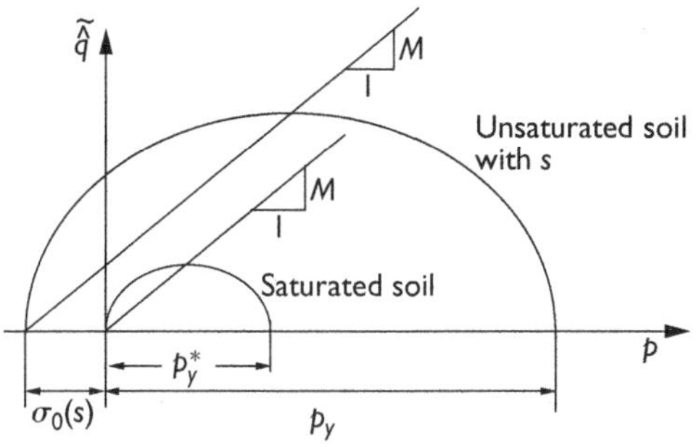

Figure 4.3.11 Yield curves under constant suction.

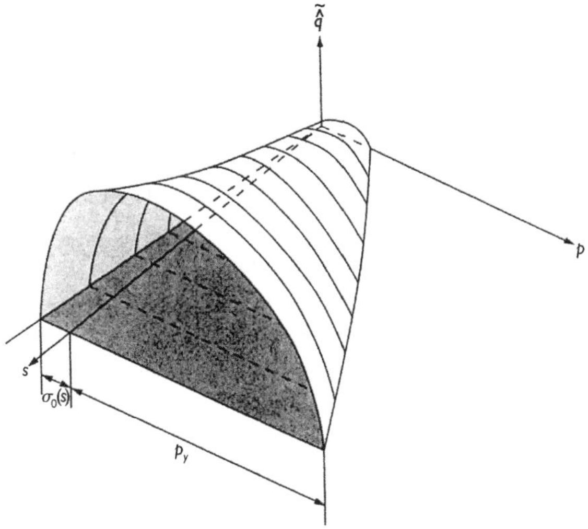

Figure 4.3.12 Yield surface for unsaturated soil.

where the proportionality constant Λ can be determined from the consistency condition. Equation (4.3.32) can be rewritten as $f = f(\hat{p}, \hat{q}, p_y, \sigma_0(s)) = f(\sigma_{ij}, p_y, s) = 0$, so we have

$$df = \frac{\partial f}{\partial \sigma_{ij}} d\sigma_{ij} + \frac{\partial f}{\partial p_y} dp_y + \frac{\partial f}{\partial s} ds = 0 \tag{4.3.34}$$

Substituting Eq. (4.3.29) into Eq. (4.3.34) and arranging it gives

$$df = \frac{\partial f}{\partial \sigma_{ij}} d\sigma_{ij} + \frac{\partial f}{\partial p_y} \frac{\partial p_y}{\partial p_y^*} dp_y^* + \left(\frac{\partial f}{\partial p_y} \frac{\partial p_y}{\partial s} + \frac{\partial f}{\partial s} \right) ds = 0 \tag{4.3.35}$$

in which the isotropic yielding stress p_y^* for saturated soil is related to the volumetric strain ε_v^p, the same as that used in the Cam-clay model. Because the plastic volumetric strain ε_v^p is a hardening parameter in the model, $d\varepsilon_v^p$ of saturated soil induced by dp_y^* is the same as $d\varepsilon_v^p$

of unsaturated soil induced by dp_y and/or $d\sigma_0(s)$. Allowing for Eqs (4.3.20) and (4.3.33), we have

$$dp_y^* = \frac{1+e(0)}{\lambda(0)-\kappa} p_y^* d\varepsilon_v^p = \frac{1+e(0)}{\lambda(0)-\kappa} p_y^* \Lambda \frac{\partial f}{\partial \tilde{\sigma}_{ij}} \delta_{ij} \tag{4.3.36}$$

Substituting Eq. (4.3.36) into Eq. (4.3.35) and solving for Λ gives

$$\Lambda = -\frac{(\partial f/\partial \sigma_{ij})d\sigma_{ij} + ((\partial f/\partial p_y)(\partial p_y/\partial s) + (\partial f/\partial s))ds}{(\partial f/\partial p_y)(\partial p_y/\partial p_y^*)p_y^*(1+e_i(0))/(\lambda(0)-\kappa)(\partial f/\partial \tilde{\sigma}_{ij})\delta_{ij}} \tag{4.3.37}$$

When $\sigma_0(s) = 0$, the proportionality constant Λ for unsaturated soil (Eq. (4.3.37) becomes that of the modified Cam-clay model revised by the SMP criterion as described in Chapter 3, i.e.,

$$\Lambda = -\frac{(\partial f/\partial \sigma_{ij})d\sigma_{ij}}{(\partial f/\partial p_y^*)p_y^*(1+e_i(0))/(\lambda(0)-\kappa)(\partial f/\partial \tilde{\sigma}_{ij})\delta_{ij}} \tag{4.3.38}$$

Therefore, the present elastoplastic model for unsaturated soil includes the modified Cam-clay model for saturated soil as a special case.

From Eqs (4.3.33) and (4.3.37), we can calculate the plastic strain increments induced by the increment in stress and/or the decrease in suction. The gradients $\partial f/\partial \tilde{\sigma}_{ij}$, $\partial f/\partial p_y$, and $\partial f/\partial s$ can be calculated by differentiating Equation (4.3.32), and $\partial p_y/\partial p_y^*$ and $\partial p_y/\partial s$ are given in Eqs (4.3.30) and (4.3.31). $\partial f/\partial \sigma_{ij}$ can be calculated using the derivative of compound function.

$$\frac{\partial f}{\partial \sigma_{ij}} = \frac{\partial f}{\partial \tilde{p}} \frac{\partial \tilde{p}}{\partial \sigma_{ij}} + \frac{\partial f}{\partial \tilde{q}} \frac{\partial \tilde{q}}{\partial \sigma_{ij}} \tag{4.3.39}$$

4.3.5 Triaxial tests on unsaturated soils

The elastoplastic model for unsaturated soils has been explained in the previous sections. However, it is necessary to validate whether the model can describe the main deformation and strength characteristics of unsaturated soils. In order to do this, triaxial tests on unsaturated soil were carried out.

4.3.5.1 Triaxial test apparatus for unsaturated soils

Tests were conducted in an improved triaxial apparatus, which can directly measure the lateral strain of specimens and control suction. The lateral displacement is measured by using two rings made of bronze, as shown in Figure 4.3.13, which are set at $\frac{1}{4}$ and $\frac{1}{2}$ heights of specimens from the top. The axis translation technique is adopted to control the suction. In the triaxial cell, the pore-water pressure is maintained to be atmospheric during the test

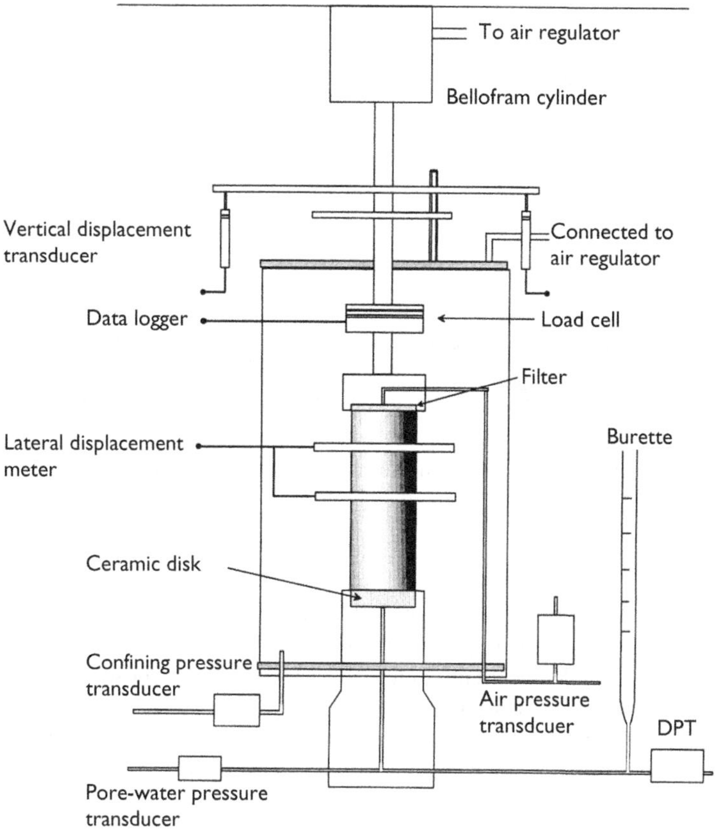

Figure 4.3.13 Triaxial apparatus for testing unsaturated soils.

through a ceramic disk, which is installed in the pedestal, with the air entry value of 300 kPa, while the air pressure is applied at the top through a polyfuortetraethylene filter.

4.3.5.2 Unsaturated soil specimen

Triaxial specimens, 3.5 cm in diameter and 8.0 cm high, were prepared by compaction in a mould at water content of 26%. The soil used is called "Pearl-clay." It contains 50% silt and 50% clay, and has a liquid limit of 49% and a plasticity index of 22. All specimen were compacted in five layers, with each layer statically compacted 15 times using a bar up to a vertical stress of 314 kPa. This procedure results in a dry density of about 1.20 g/cm^3, a void ratio of about 1.30, a degree of saturation of about 50%, and an initial suction of about 95 kPa. All specimens were compacted in the same way for producing the almost same initial soil fabric in every test. For further details, see Sun *et al.* (2004).

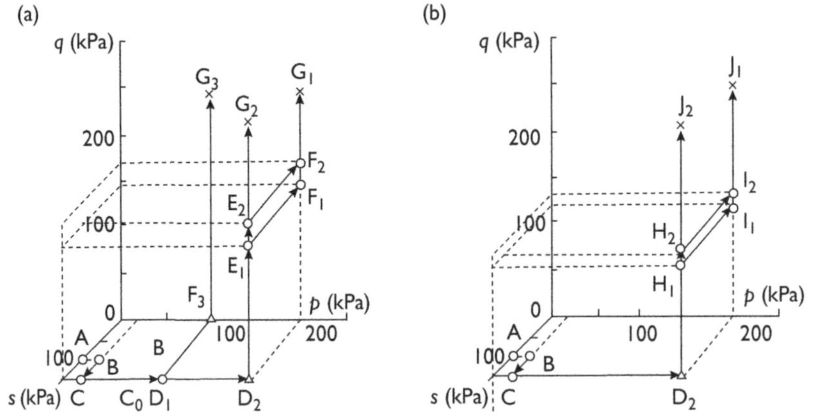

Figure 4.3.14 Stress paths in (a) triaxial compression and (b) triaxial extension.

4.3.5.3 Stress paths

The specimens were first isotropically consolidated at a mean stress of 20 kPa, then an air pressure of 147 kPa was applied to specimens with $u_w = 0$ and $p = 20$ kPa, and the specimens were loaded to a specified stress state. Figure 4.3.14 shows the imposed stress paths, where point A represents the initial state with the initial suction of about 95 kPa produced by the compaction process. Table 4.3.1 shows the stress path of every test in detail. After the isotropic consolidation, the triaxial compression and extension tests on unsaturated compacted Pearl-clay were conducted. The tests include the collapse tests performed by decreasing suction under isotropic and anisotropic stress states. For further details of the collapse tests see Sun *et al.* (2004).

4.3.6 Comparison of model predictions with experimental results

4.3.6.1 Model parameters and their determination

Because the proposed constitutive relation is formulated within the elastoplastic framework, the strains consist of elastic and plastic components. The model requires six material parameters $\lambda(0), \lambda_s, \kappa, M(0), M_s$, and a, and the initial state values $p_i, e_i(0)$, and $e_i(s)$. These model parameters are determined from isotropic compression tests with loading-unloading-reloading process and subsequent triaxial compression tests on saturated and unsaturated soils under constant suction and constant p or constant confining net stress. In detail, $\lambda(0), \lambda_s, \kappa, p_i, e_i(0)$, and $e_i(s)$ are determined from the results of isotropic compression tests on saturated and unsaturated soils with loading-unloading process, as shown in Figure 4.3.8(a), whereas $M(0)$ and M_s are determined from the envelope lines of Mohr's circles at failure measured by a triaxial compression test on saturated soil and two triaxial compression tests on unsaturated soil under constant suction at different confining net stress, using Eq. (4.3.18); a is determined as shown in Figure 4.3.3 from s and corresponding $\sigma_0(s)$,

Table 4.3.1 Stress path of every test

Kinds of test	No	*Consolidation*	*Shear*
Isotropic consolidation	Path I	$A \rightarrow B \rightarrow C(s = 147) \rightarrow D_1(p = 98) \rightarrow F_3(s = 0)$	—
Triaxial compression ($s = 0$)	Path II	$A \rightarrow B \rightarrow C(s = 147) \rightarrow D_1 \rightarrow F_3(p = 98)$	$F_3(s = 0, p = 98) \rightarrow G_3$
Triaxial compression ($s =$ const.)	Path III	$A \rightarrow B \rightarrow C(s = 147) \rightarrow D_2(p = 196)$	$D_2(p = 196) \rightarrow G_2$
Triaxial extension ($s =$ const.)	Path IV	$A \rightarrow B \rightarrow C(s = 147) \rightarrow D_2(p = 196)$	$D_2(p = 196) \rightarrow J_2$
Triaxial compression	Path V	$A \rightarrow B \rightarrow C(s = 147) \rightarrow D_2(p = 196)$	$D_2(p = 196) \rightarrow E_1(q = 147) \rightarrow F_1(s = 0) \rightarrow G_1$
(Wetting at shear)	Path VI	$A \rightarrow B \rightarrow C(s = 147) \rightarrow D_2(p = 196)$	$D_2(p = 196) \rightarrow E_2(q = 172) \rightarrow F_2(s = 0) \rightarrow G_1$
Triaxial extension	Path VII	$A \rightarrow B \rightarrow C(s = 147) \rightarrow D_2(p = 196)$	$D_2(p = 196) \rightarrow H_1(q = 118) \rightarrow I_1(s = 0) \rightarrow J_1$
(Wetting at shear)	Path VII	$A \rightarrow B \rightarrow C(s = 147) \rightarrow D_2(p = 196)$	$D_2(p = 196) \rightarrow H_2(q = 132) \rightarrow I_2(s = 0) \rightarrow J_1$

Note
Unit of p, q, and s is kPa.

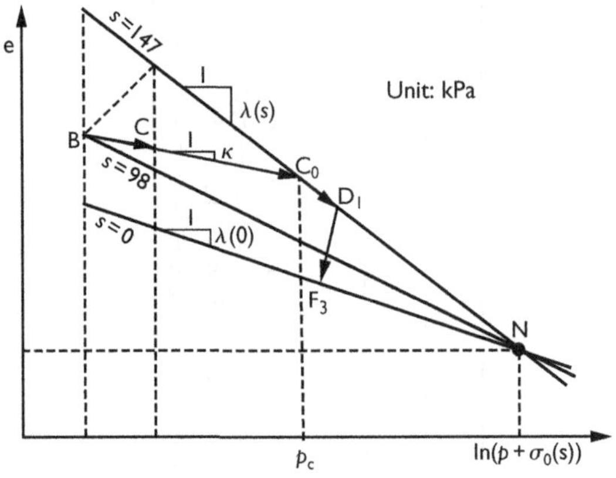

Figure 4.3.15 Model prediction for variation of void ratio with p along isotropic stress path $\mathrm{B} \rightarrow \mathrm{C} \rightarrow \mathrm{C}_0 \rightarrow \mathrm{D}_1 \rightarrow \mathrm{F}_3$.

which is the distance between the origin and the intersection point of the σ-axis and the envelope line of measured Mohr's circles at failure.

The model parameters could be also determined approximately from the results of the suction-controlled oedometer and direct shear tests. Parameters $\lambda(0), \lambda_s, \kappa, p_i, e_i(0)$, and $e_i(s)$ are determined from the results of the suction-controlled oedometer tests on saturated and unsaturated soils with loading and unloading process, and $M(0), M_s$, and a are determined from the results of a direct shear test on saturated soil and two suction-controlled direct shear tests on unsaturated soil under constant suction and different vertical stresses.

The elastic component is calculated from Hooke's law, while Poisson's ratio is assumed to be zero, and the elastic modulus is calculated from that in the same way as the Cam-clay model, i.e.,

$$E = \frac{3\hat{p}(1 + e_i(s))}{\kappa} \tag{4.3.40}$$

The values of the relevant model parameters used in predicting the stress–strain behavior of tested unsaturated Pearl-clay are summarized in Table 4.3.2.

4.3.6.2 Model predictions versus experimental results

Figure 4.3.15 shows the model prediction for the variation of the void ratio(e) with the mean net stress (p) along the isotropic stress path I ($\mathrm{B} \rightarrow \mathrm{C} \rightarrow \mathrm{C}_0 \rightarrow \mathrm{D}_1 \rightarrow \mathrm{F}_3$; see Fig. 4.3.14(a) and Table 4.3.1. Here C_0 is the yielding point with $p = p_c$ at $s = 147$ kPa) which involves the increment in suction from the initial suction ($s_0 = 95$ kPa) to $s = 147$ kPa ($\mathrm{B} \rightarrow \mathrm{C}$: $s = 95 \rightarrow 147$ kPa at $p = 20$ kPa), subsequent increment in the mean net stress ($\mathrm{C} \rightarrow \mathrm{D}_1$: $p = 20 \rightarrow 98$ kPa at $s = 147$ kPa), and the reduction in suction ($\mathrm{D}_1 \rightarrow \mathrm{F}_3$: $s = 147 \rightarrow 0$ kPa at $p = 98$ kPa). In Figure 4.3.15 and the following Figures 4.3.16–4.3.20, signs A, D_1, and H_1, for example, correspond to the same signs in Figure 4.3.14. In Figure 4.3.15, respectively,

Table 4.3.2 Material parameters and initial state values

Material parameters							*Initial state values*		
M(0)	*Ms (1/kPa)*	λ(0)	λ_s *(1/kPa)*	κ	ν	*a(kPa)*	p_i*(kPa)*	*s(kPa)*	$e_i(s)$
1.05	0.00625	0.10	0.0015	0.03	0.33	53	98	147 0	127 1.11

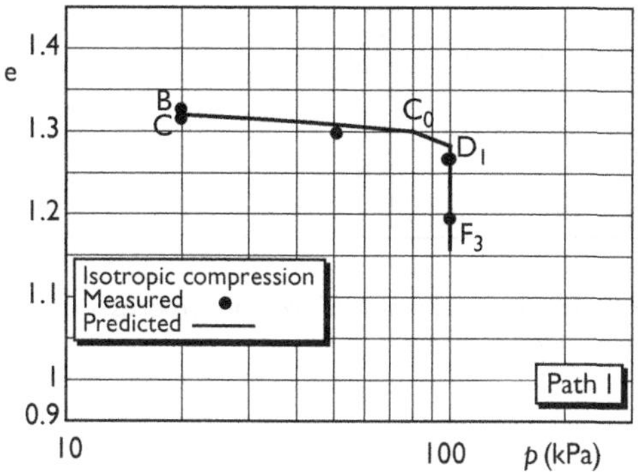

Figure 4.3.16 Predicted and experimental result of isotropic compression test including decreasing suction.

the stress path B→C is in the elastic region with increasing suction ($s = 95 \rightarrow 147$ kPa) at constant mean net stress ($p = 20$ kPa), the corresponding change of the void ratio is assumed to be related to the change in the suction stress $\sigma_0(s)$ through the elastic swelling index κ. The stress path C → C_0 is also in the elastic region with increasing mean net stress ($p = 20$ kPa $\rightarrow p_c$) at constant suction ($s = 147$ kPa), the corresponding change of the void ratio can be related to the change in the mean net stress by using the elastic swelling index κ. The stress path C_0 →D_1 is in the normal compression region with increasing mean net stress ($p = p_c \rightarrow 98$ kPa) at constant suction ($s = 147$ kPa), the corresponding change of the void ratio can be related to the change in the mean net stress through the compression index $\lambda(s)$ (Eq. (4.3.23)). Along the stress path D_1 →F_3, in the suction decreases under constant mean net stress of 98 kPa, the corresponding change of the void ratio (collapse) can be calculated by Eq. (4.3.24).

Figure 4.3.16 shows the comparison between the measured and predicted results of unsaturated compacted Pearl-clay under the conditions of increasing suction and constant mean net stress (B →C), increasing mean net stress and constant suction (C→ C_0 →D_1), and constant mean net stress and decreasing suction (D_1 →F_3), all under isotropic stress states. As described in the previous few lines, the unsaturated soil behaves elastically in the stress path B → C → C_0, i.e., Hooke's law is adopted while the elastic modulus is calculated from κ as expressed in Eq. (4.3.40) regardless of the suction increase (B→C) or the mean net stress increase (C→C_0).

Figure 4.3.17 shows the comparison between the measured and predicted results of triaxial compression tests on saturated compacted Pearl-clay under constant mean effective stress. In

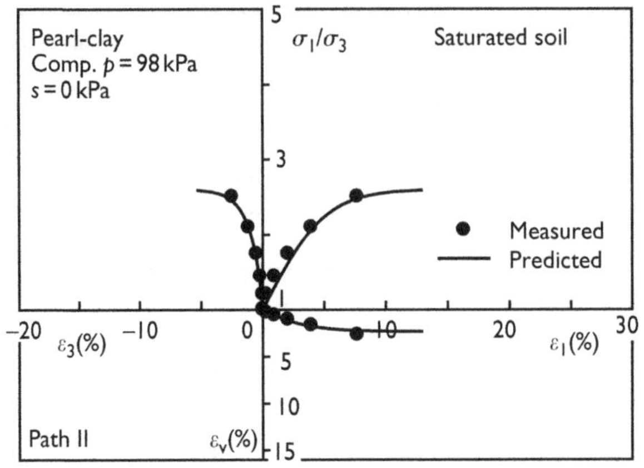

Figure 4.3.17 Predicted and experimental result of triaxial compression test on saturated soil under constant mean effective stress.

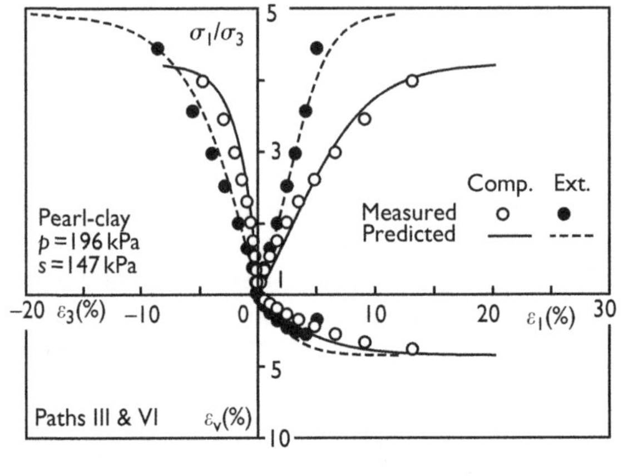

Figure 4.3.18 Predicted and experimental results of triaxial compression and extension tests on unsaturated soil under constant suction and constant mean net stress.

Figure 4.3.17 and the following figures, σ_1/σ_3 denotes the major-minor principal stress ratio, and ε_1, ε_3, and ε_v denote the major, minor principal strains and volumetric strain, respectively. Figure 4.3.18 shows the comparison between the measured and predicted results of triaxial compression and extension tests on unsaturated compacted Pearl-clay under constant mean net stress (p = 196 kPa) and constant suction (s = 147 kPa). It can be seen from Figures 4.3.17 and 4.3.18 that the model can predict accurately the mechanical behavior of saturated and unsaturated soils under constant suction in triaxial compression and triaxial extension stresses. Figures 4.3.19 and 4.3.20 show the comparison between the measured and predicted results of triaxial compression and extension tests on unsaturated compacted

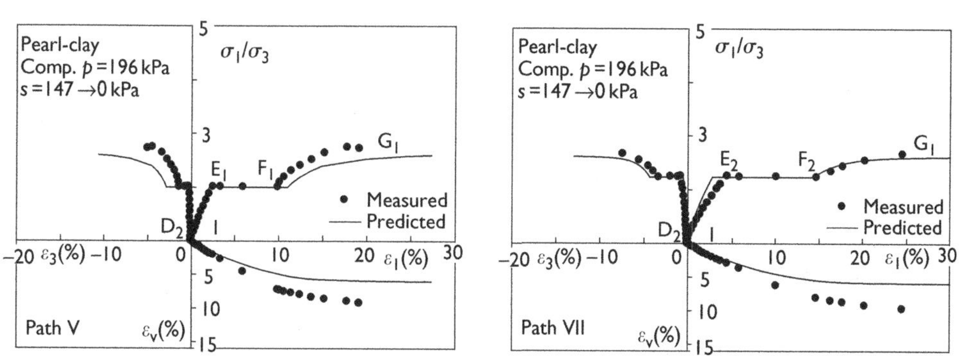

Figure 4.3.19 Predicted and experimental results of triaxial compression tests on unsaturated soil under $p = 196$ kPa and decreasing suction ($s = 147$ kPa $\rightarrow$ 0 kPa) at different stress ratios during shearing.

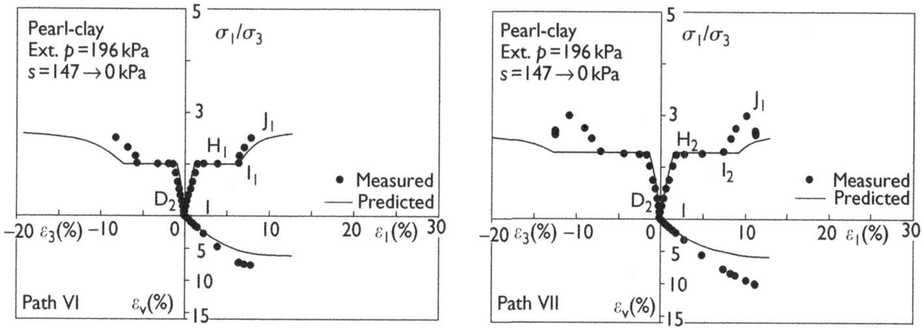

Figure 4.3.20 Predicted and experimental results of triaxial extension tests on unsaturated soil under $p = 196$ kPa and decreasing suction ($s = 147$ kPa $\rightarrow$ 0 kPa) at different stress ratios during shearing.

Pearl-clay under constant mean net stress (p), in which stress paths include suction reduction ($s = 147$ kPa$\rightarrow$0 kPa) at different stress ratios during shearing. It can be seen from the measured results shown in Figures 4.3.19 and 4.3.20 that the shear strain (i.e., $\varepsilon_1 - \varepsilon_3$) increment caused by collapse at high stress ratio (σ_1/σ_3) ($E_2 \rightarrow F_2$ in Fig. 4.3.19 and $H_2 \rightarrow I_2$ in Fig. 4.3.20) is greater than that at low stress ratio ($E_1 \rightarrow F_1$ in Fig. 4.3.19 and $H_1 \rightarrow I_1$ in Fig. 4.3.20), while the volumetric strain increment is almost the same at the two tested stress ratio. This phenomenon shows that the volumetric strain increment caused by collapse from metastable unsaturated soil to stable saturated soil is mainly dependent on the mean stress, while the shear strain increment is mainly dependent on the ratio of shear stress, at which suction decreases, to shear strength of saturated soil. The model describes well this behavior, i.e., the shear strain increment induced by decreasing suction at high stress ratio ($E_2 \rightarrow F_2$ in Fig. 4.3.19 and $H_2 \rightarrow I_2$ in Fig. 4.3.20) is greater than that at low stress ratio ($E_1 \rightarrow F_1$ in

Fig. 4.3.19 and $H_1 \rightarrow I_1$ in Fig. 4.3.20), while the volumetric stain increment is almost the same at the two-tested stress ratio. It is interesting to note that the model can predict well the whole stress–strain relation of unsaturated soil during the process of constant suction ($D_2 \rightarrow E_1$ and $D_2 \rightarrow E_2$ in Fig. 4.3.19 and $D_2 \rightarrow H_1$ and $D_2 \rightarrow H_2$ in Fig. 4.3.20), the suction reduction ($E_1 \rightarrow F_1$and $E_2 \rightarrow F_2$ in Fig. 4.3.19 and $H_1 \rightarrow I_1$and $H_2 \rightarrow I_2$ in Fig. 4.3.20), and zero suction ($F_1 \rightarrow G_1$and $F_2 \rightarrow G_1$ in Fig. 4.3.19 and $I_1 \rightarrow J_1$ and $I_2 \rightarrow J_1$ in Fig. 4.3.20) in triaxial compression and triaxial extension.

It can be concluded from Figures 4.3.16–4.3.20 that the proposed model predicts well the stress–strain behavior including collapse of unsaturated soil under the isotropic or general stress state, using the compression and swelling indexes and strength parameters of saturated and unsaturated soils.

4.3.7 General stress–strain relation for unsaturated soils

The elastoplastic constitutive tensor of the present model for unsaturated soils used in solving elastoplastic boundary value problem by means of the finite element method is derived here. The elastic part of the stress–strain relation can be written in the incremental form as

$$d\sigma_{ij} = D^e_{ijkl} d\varepsilon^e_{kl} = D^e_{ijkl}(d\varepsilon_{kl} - d\varepsilon^p_{kl}) \tag{4.3.41}$$

Where D^e_{ijkl} is the elastic constitutive tensor. ε_{kl} and ε^p_{kl} are the total strain tensor and its plastic component respectively.

Substituting Eqs (4.3.41), (4.3.33), and (4.3.36) into Eq. (4.3.35) gives

$$\Lambda = \frac{\frac{\partial f}{\partial \sigma_{ij}} D^e_{ijkl} d\varepsilon_{kl} + \left(\frac{\partial f}{\partial p_y}\frac{\partial p_y}{\partial s} + \frac{\partial f}{\partial s}\right) ds}{X} \tag{4.3.42}$$

where

$$X = \frac{\partial f}{\partial \sigma_{ij}} D^e_{ijkl} \frac{\partial f}{\partial \tilde{\sigma}_{kl}} - \frac{1+e}{\lambda(0) - \kappa} p^*_y \frac{\partial f}{\partial p_y}\frac{\partial p_y}{\partial p^*_y}\frac{\partial f}{\partial \tilde{\sigma}_{ij}} \delta_{ij} \tag{4.3.43}$$

Substituting Eqs (4.3.33) and (4.3.42) into Eq. (4.3.41), we get a general form of the 'stress'–strain relation in the incremental form.

$$d\sigma_{ij} = D_{ijkl} d\varepsilon_{kl} + W^{ep}_{ij} ds \tag{4.3.44}$$

where

$$D_{ijkl} = D^e_{ijkl} - D^e_{ijmn} \frac{\partial f}{\partial \tilde{\sigma}_{mn}} \frac{\partial f}{\partial \sigma_{st}} D^e_{stkl} / X \tag{4.3.45}$$

$$W^{ep}_{ij} = \left(\frac{\partial f}{\partial p_y}\frac{\partial p_y}{\partial s} + \frac{\partial f}{\partial s}\right) D_{ijkl} \frac{\partial f}{\partial \tilde{\sigma}_{kl}} / X \tag{4.3.46}$$

4.3.8 Concluding remarks

1 An effective stress for unsaturated soil is expressed as $\hat{\sigma}_{ij} = \sigma_{tij} - u_a\delta_{ij} + \sigma_0(s)\delta_{ij}$. The strength criterion and the present elastoplastic model for unsaturated soil in three-dimensional stresses are formulated using this effective stress, to account for some of the suction effects on the mechanical behavior of unsaturated soil.

2 A three-dimensional elastoplastic constitutive model for unsaturated soil has been introduced. The model is extended from the elastoplastic model for cohesive-frictional materials with constant cohesion by using the relationship between deformation characteristics of saturated and unsaturated soils in the isotropic stress, which accounts for the other suction effects on the mechanical behavior of unsaturated soil.

3 Various kinds of tests on unsaturated soil using a suction-controlled triaxial apparatus were conducted, especially triaxial compression and extension tests with the collapse (wetting) caused by decreasing suction at different stress ratios. Comparison of the model prediction and the experimental results shows that the present model predicts well the stress–strain–strength behavior including collapse of unsaturated soil in three-dimensional stresses.

The model described in this section is essentially suitable to predict the mechanical behavior of unsaturated soils with one initial density. That is to say, the model parameters depend on the initial density. As well known, the initial density of compacted soil varies with compaction. So, even for the same soil, the values of the model parameters vary with the initial density, which means that the model parameters are not the material parameters. In order to avoid this contradiction, it is necessary to account for the initial density dependency of unsaturated soil behavior. For details about how to account for this dependency in the model refer to Sun *et al.* (2003).

Another shortcoming of the present model is that the degree of saturation is computed from the suction using the water retention curve, which cannot take into account the influence of the saturation degree on the mechanical behavior of unsaturated soils. For details about how to account for this influence in the model refers to Sheng *et al.* (2004) and Sun and Sheng (2005).

References

Bishop, A.W. 1959. Factors controlling the strength of partly saturated cohesive soils. Proceedings of Shear strength of Cohesive Soils, ASCE, Colorado, pp. 503–532.

Dafalias, Y.F. 1986. Bounding surface plasticity. I: Theory. *Journal of Engineering Mechanics ASCE*, 112(12): 1242–1291.

Escario, V. and Saez, J. 1987. Shear strength of soils under high suction values. Session 5; Proceedings of the 9th European Conference on Soil Mechanics, Vol. 3, Written discussion, p. 1157.

Escario, V. and Juca, J.F.T. 1989. Shear and deformation of partly saturated soils. Proceedings of the 12th ICSMFE, Vol. 1, pp. 43–46.

Fredlund, D.G., Morgenstern, N.R., and Widger, R.S. 1978. The shear strength of unsaturated soils. *Canadian Geotechnical Journal*, 15(3): 313–321.

Fredlund, D.G., Rahardjo H., and Gan J.K.M. 1987. Nonlinearity of strength envelope for unsaturated soils. Proceedings of the 6th International Conference on Expansive Soils, pp. 49–54.

Funada, T., Matsuoka, H., and Fukumoto, S. 1989. Deformation characteristics of sand under various stress paths from anisotropic consolidation. Proceedings of the 44th Annual Conference of the Japan Society of Civil Engineers, Vol. 3, pp. 512–513 (in Japanese).

Hashiguchi, K. and Ueno, M. 1977. Elasto-plastic constitutive laws of glandular materials. Proceedings of the Special Session 9 of 9th International Conference on Soil Mechanics and Foundations Engineering, Tokyo, pp. 73–82.

Hata, S., Ohta, H., and Yoshitami, S. 1969. On the state surface of soils. *Proceedings of JSCE*, 172: 97–117.

Ichihara, W. 1997. Deformation and strength characteristics of intermediate soil under triaxial stress condition. Bachelor's thesis at Nagoya Institute of Technology (in Japanese).

Ishihara, K., Tatsuoka, F., and Yasuda, S. 1975. Undrained deformation and liquefaction of sand under cyclic stresses. *Soils and Foundations*, 15(1): 29–44.

Lade, P.V. 1977. Elasto-plastic stress-strain theory for cohesionless soils with curved yield surface. *International Journal of Solids and Structures*, 13, 1019–1035.

Lade, P.V. and Duncan, J.M. 1975. Elasto-plastic stress-strain theory for cohesionless soil. *Journal of Geotechnical Engineering ASCE,* 101(10): 1037–1053.

Matsuoka, H. and Nakai, T. 1974. Stress-deformation and strength characteristics of soil under three different principal stresses. *Proceedings of JSCE*, 232. 59–74.

Matsuoka, H. and Sun, D.A. 1995. Extension of spatially mobilized plane (SMP) to frictional and cohesive materials and its application to cemented sands. *Soils and Foundations*, 35(4): 63–72.

Matsuoka H., Hoshikawa T., and Ueno K. 1990. A general failure criterion and stress-strain relation for granular materials to metals. *Soils and Foundations*, 30(2): 119–127.

Matsuoka, H., Yao, Y.P., and Sun, D.A. 1999. The Cam-clay models revised by the SMP criterion. *Soils and Foundations*, 39(1): 81–95.

Matsuoka H., Sun D.A., Kogane A., Fukuzawa N., and Ichihara W. 2002. Stress-strain behaviour of unsaturated soil in true triaxial tests. *Canadian Geotechnical Journal*, 39(3): 608–619.

Nakai, T. 1989. An isotropic hardening elastoplastic model for sand considering the stress path dependency in three-dimensional stresses. *Soils and Foundations*, 29(1): 19–137.

Nakai, T. and Matsuoka, M. 1986. A generalized elastoplastic constitutive model for clay in three-dimensional stresses. *Soils and Foundations*, 26(3): 81–98.

Nakai, T., Tsuzuki, K., Yamamoto, M., and Hishida, T. 1986. Analysis of plane strain tests on normally consolidated clay by an elastoplastic constitutive model. Proceedings of the 21st Japan National Conference on Soil Mechanics and Foundation Engineering, Vol. 2, pp. 453–456 (in Japanese).

Nakai, T., Tsuzuki, K., Ishikawa, K., and Miyake, M. 1987. Analysis of plane strain tests on normally consolidated clay by an elastoplastic constitutive model. Proceedings of the 22nd Japan National Conference on Soil Mechanics and Foundation Engineering, Vol. 1, pp. 419–420 (in Japanese).

Nova, R. and Wood, D.M. 1979. A constitutive model for sand in triaxial compression. *International Journal Numerical and Analytical methods in Geomechanics*, 3(3): 255–278.

Ohta, H. and Hata, S. 1971. On the state surface of anisotropically consolidated clays. *Proceedings of JSCE*, 196: 117–124.

Ohta, H. and Nishihara, A. 1985. Anisotropy of undrained shear strength of clays under axi-symmetric loading conditions. *Soils and Foundations*, 25(2): 73–86.

Pastor, M., Zienkiewicz, O.C., and Leung, KH. 1985. A simple model for transient loading in earthquake analysis. part II: Non-associative model for sands. *International Journal for Numerical and Analytical Methods in Geomechanics*, 9: 477–498.

Prevost, J.H. 1985. A simple plasticity theory for frictional cohesionless soils. *Soil Dynamics and Earthquake Engineering*, 4(1): 9–17.

Roscoe, K.H. and Burland, J.B. 1968. On the generalised stress-strain behaviour of "wet" clay. *Engineering Plasticity*, Cambridge University Press, pp. 535–609.

Roscoe, K.H., Schofield, A.N., and Thurairajah, A. 1963. Yielding of clay in state wetter than critical. *Geotechnique*, 13(3): 211–240.

Sekiguchi, H. and Ohta, H. 1977. Induced anisotropy and time dependency in clays. Proceedings of the 9th ICSMFE, Specialty Session 9, pp. 229–238.

Sheng, D.C., Sloan, S.W., and Gens, A. 2004. A constitutive model for unsaturated soils: thermomechanical and algorithmic aspects. *Computational Mechanics*, 33: 453–465.

Sun, D.A., and Sheng, D.C. 2005. An elastoplastic hydro-mechanical model for unsaturated compacted soils. Proceedings International Symposium on Advanced Experimental Unsaturated Soil Mechanics, Trento, pp. 254–260.

Sun, D.A., Matsuoka, H., Yao, Y.P., and Ichihara, W. 2000. An elasto-plastic model for unsaturated soil in three-dimensional stresses. *Soils and Foundations*, 40(3): 17–28.

Sun, D.A., Matsuoka, H., Yao, Y.P., and Ichimura, M. 2001. A transformed stress based on extended SMP criterion and its application to elastoplastic model for geomaterials. *Journal of Geotechnical Engineering*, Japan Society of Civil Engineers (JSCE), No. 680/III-55: 211–224.

Sun, D.A., Matsuoka, H., Cui, H.B., and Xu, Y.F. 2003. Three-dimensional elasto-plastic model for unsaturated compacted soils with different initial densities. *International Journal for Numerical and Analytical Method in Geomechanics*, 27(12): 1079–1098.

Sun, D.A., Matsuoka, H., and Xu, Y.F. 2004. Collapse behavior of compacted clay by suction-controlled triaxial tests. *Geotechnical Testing Journal ASTM*, 27(4): 362–370.

Sun, D.A., Matsuoka, H., Yao, Y.P., and Ishii, H. 2004. Anisotropic hardening elastoplastic model for clays and sands and its application to FE analysis. *Computers and Geotechnics*, 31(1): 37–46.

Terzaghi, K. 1936. The shear resistance of saturated soils and the angles between the planes of shear. Proceedings of the 1st International Conference on Soil Mechanics and Foundation Engineering, Boston, Vol. 1, pp. 54–56.

Wroth, C.P. and Houlsby, G.T. 1985. Soil mechanics-property characterization and analysis procedures. Proceedings of 11th International Conference on Soil Mechanics and Foundations Engineering, San Francisco, Vol. 1, pp. 1–55.

Yamada, K. 1979. Deformation characteristics of loose sand under three-dimensional stress. Doctoral thesis at University of Tokyo (in Japanese).

Yao, Y.P., Matsuoka, H., and Sun, D.A. 1999. A unified elastoplastic model for clay and sand with the SMP criterion. Proceedings of the 8th Australia-New Zealand Conference on Geomechanics, Hobart, Vol. 2, pp. 997–100.

Zienkiewicz, O.C., Leung, K.H., and Pastor, M. 1985. A simple model for transient loading in earthquake analysis. Part I: Basic Model. International Journal for Numerical and Analytical Methods in Geomechanics, Vol. 9, pp. 453–476.

Chapter 5

Concluding remarks

In this book, we first presented the original concepts of the Spatially Mobilized Plane (SMP) and the SMP criterion, which is an extension of the Mohr-Coulomb criterion to three-dimensional stress states. The Cam-clay models were then introduced briefly and the combination of the SMP criterion and the Cam-clay models were achieved using a transformed stress $\tilde{\sigma}_{ij}$. Several elastoplastic models for geomaterials were then presented using the transformed stress $\tilde{\sigma}_{ij}$ and the SMP criterion. These models are applicable to sands exhibiting positive and negative dilation, sands and clays with K_0-consolidation anisotropy, and unsaturated soils. The elastoplastic constitutive tensors of these models for solving boundary value problems by means of the finite element method were also given. Table 5.1 shows a list of the models introduced in this book. This table also gives the mutual relationships between the Cam-clay model and our models.

Table 5.1 A list of the introduced elastoplastic models for geomaterials

① *Model*	② *Soil*	③ *Stress state*	④ *Stress-dilatancy equation*	⑤ *Plastic potential and yield function*	⑥ *Flow rule*	⑦ *Hardening parameter*	⑧ *Hardening law*	⑨ *Failure criterion*
Cam-clay model (CCM)	Normally consolidated clay	Triaxial compression	$\frac{q}{p} = M - \frac{d\varepsilon_v^p}{d\varepsilon_d^p}$	$g = f = \frac{\lambda - \kappa}{1 + e_0} \times \left[\ln \frac{p}{p_0} + \frac{1}{M}\frac{q}{p}\right] - \varepsilon_v^p = 0$	$d\varepsilon_{ij} = \Lambda \frac{\partial f}{\partial \sigma_{ij}}$	ε_v^p	$\varepsilon_v^p = \frac{\lambda - \kappa}{1 + e_0} \times \ln \frac{p_y}{p_0}$	$\frac{q}{p} = M$
CCM-SMP	Normally consolidated clay	Three-dimensional stress	$\frac{\tilde{q}}{\tilde{p}} = M - \frac{d\varepsilon_v^p}{d\varepsilon_d^p}$	$g = f = \frac{\lambda - \kappa}{1 + e_0} \times \left[\ln \frac{\tilde{p}}{\tilde{p}_0} + \frac{1}{M}\frac{\tilde{q}}{\tilde{p}}\right] - \varepsilon_v^p = 0$	$d\varepsilon_{ij} = \Lambda \frac{\partial f}{\partial \tilde{\sigma}_{ij}}$	ε_v^p	$\varepsilon_v^p = \frac{\lambda - \kappa}{1 + e_0} \times \ln \frac{p_y}{p_0}$	$\frac{\tilde{q}}{\tilde{p}} = M$
CCM-SMP-H	Normally consolidated sand and clay	Three-dimensional stress	$\frac{\tilde{q}}{\tilde{p}} = \sqrt{M^2 + (\frac{d\varepsilon_v^p}{d\varepsilon_d^p})^2} - \frac{d\varepsilon_v^p}{d\varepsilon_d^p}$	$g = f = \frac{\lambda - \kappa}{1 + e_0} \times \left[\ln \frac{\tilde{p}}{\tilde{p}_0} + \ln\left(1 + \frac{\tilde{q}^2}{M^2\tilde{p}^2}\right)\right] - \tilde{H} = 0$	$d\varepsilon_{ij} = \Lambda \frac{\partial f}{\partial \tilde{\sigma}_{ij}}$	$d\tilde{H} = \frac{M^4}{M_f^4} \times \frac{M_f^4 - \tilde{\eta}^4}{M^4 - \tilde{\eta}^4} d\varepsilon_v^p$	$\varepsilon_v^p = \frac{\lambda - \kappa}{1 + e_0} \times \ln \frac{p_y}{p_0}$	$\frac{\tilde{q}}{\tilde{p}} = M_f$
CCM-SMP-H taking K_0-consolidation into account	K_0-consolidated sand and clay	Three-dimensional stress	$\tilde{\eta}_k = M - \frac{d\varepsilon_v^p}{d\varepsilon_d^p}$	$g = f = \frac{\lambda - \kappa}{1 + e_0} \times \left[\ln \frac{\tilde{p}}{\tilde{p}_0} + \frac{\tilde{\eta}^*}{M}\right] - \tilde{H} = 0$	$d\varepsilon_{ij} = \Lambda \frac{\partial f}{\partial \tilde{\sigma}_{ij}}$	$d\tilde{H} = \frac{M^4}{M_f^4} \times \frac{M_f^4 - \tilde{\eta}_k^4}{M^4 - \tilde{\eta}_k^4} d\varepsilon_v^p$	$\varepsilon_v^p = \frac{\lambda - \kappa}{1 + e_0} \times \ln \frac{p_y}{p_0}$	$\tilde{\eta}_k = M_f$
CCM-SMP for change in σ_0	Unsaturated soil	Three-dimensional stress	$\frac{\tilde{\hat{q}}}{\tilde{\hat{p}}} = M(s) - \frac{d\varepsilon_v^p}{d\varepsilon_d^p}$	$g = f = \frac{\lambda(s) - \kappa}{1 + e_0(s)} \times \left[\ln \frac{\tilde{\hat{p}}}{p_0} + \frac{\tilde{\hat{q}}}{M(s)\tilde{\hat{p}}}\right] - \varepsilon_v^p = 0$	$d\varepsilon_{ij} = \Lambda \frac{\partial f}{\partial \tilde{\hat{\sigma}}_{ij}}$	ε_v^p	$\varepsilon_v^p = \frac{\lambda(s) - \kappa}{1 + e_0(s)} \times \ln \frac{p_y + \sigma_0(s)}{p_0}$	$\frac{\tilde{\hat{q}}}{\tilde{\hat{p}}} = M$

Index

For Product Safety Concerns and Information please contact our EU representative GPSR@taylorandfrancis.com Taylor & Francis Verlag GmbH, Kaufingerstraße 24, 80331 München, Germany

Printed and bound by CPI Group (UK) Ltd, Croydon, CR0 4YY

01/05/2025

01858549-0001